EASTERN TIMES

CONTENTS

Introduction	3
The Class R 4-4-0s of the North Eastern Railway	4-13
Tales of a B17 Spotter	14-25
No. 60100 *Spearmint*	26-29
Broxbourne Junction Signal Box	30-39
B1 at Banavie	40-41
The Last Shunting Horse	42-47
The 1935 LNER New Works Scheme	48-59
The Man from the Pru – Part 1	60-69
West Hartlepool (51C) in pictures	70-73
The Edinburgh Flyer	74-79
The Headshunt	80

The Transport Treasury

TIMES SERIES

EASTERN TIMES • ISSUE 2

8th February 1979 • 47014 at Liverpool Street with the 09.30 express for Norwich. Built by Brush Traction in Loughborough it entered service on 27th September 1963, outshopped as D1543 in two-tone green with small yellow warning panels, it was initially allocated to Sheffield Darnall (41A). The loco endured a nomadic life being reallocated no less than 12 times, its final shed being Stratford (30A) where it moved on 30th December 1972. Renumbered under TOPS as 47014 in February 1974 being withdrawn from service on 5th July 1991 and cut up by C. F. Booth (Rotherham) in April 1992. *Photo: Roger Geach*

© Images and design: The Transport Treasury 2023. Design and Text: Peter Sikes

ISBN: 978-1-913251-57-4

First published in 2023 by Transport Treasury Publishing Ltd., 16 Highworth Close, High Wycombe HP13 7PJ

The copyright holders hereby give notice that all rights to this work are reserved.
Aside from brief passages for the purpose of review, no part of this work may be reproduced, copied by electronic or other means, or otherwise stored in any information storage and retrieval system without written permission from the Publisher.
This includes the illustrations herein which shall remain the copyright of the copyright holder.

Copies of many of the images in EASTERN TIMES are available for purchase/download. In addition the Transport Treasury Archive contains tens of thousands of other UK, Irish and some European railway photographs.

www.ttpublishing.co.uk or for editorial issues and contributions email: tteasterntimes@gmail.com

Printed in Malta by Gutenberg Press.

INTRODUCTION

Welcome to the second issue of Eastern Times, thank you to those who purchased issue one and were kind enough to send comments to us.

In this edition we have more excellent articles from our contributors starting with the elegant 'R' (D20) class 4-4-0 designed by Wilson Wordsell of the North Eastern Railway, the class giving service for over 50 years. This is followed by one man's tale of how steam lingered on well after official withdrawal dates on his local line. We all have our favourite engines for many varied reasons, the elegant lines of A3 *Spearmint* are recalled and the reasons why it became a favourite explained.

We then move to a more modern era and a signalman's recollection of his working life at Broxbourne Junction in the late 70s and early 80s accompanied by excellent photographs of the trains he controlled. Then jumping to something completely different we feature Charlie, 'The Last Shunting Horse' who was the last of his kind to work on the railways, appropriately at Newmarket.

Moving to the capital there is a feature of how the LNER were a large part of the modernisation of the transport system needed to meet the demands of long-suffering commuters in an ever expanding metropolis before and after the Second World War.

Most of us would have heard of renowned photographer H. C. Casserley, who would take an annual two-week holiday to travel and photograph the railway system, this usually covered a large area of the country. We follow one such trip he made in 1954 accompanied by many of the photographs taken at various locations throughout the trip, there are so many photos that we have had to split this article in two with the conclusion to appear in issue 3. Finally for this issue we head north and have a pictorial feature of West Hartlepool followed by a trip on 'The Edinburgh Fast'.

As we mentioned in issue 1, we welcome your comments, and if you would like, your contributions whether that be in article or image form (or both). Please contact us at the email address below or the address found on page 2. We hope you enjoy this issue and we are already working on the next one.

PETER SIKES – EDITOR, EASTERN TIMES
email: tteasterntimes@gmail.com

Front cover (and inset right):
No. 502 *Earl Marischal* passing through North Queensferry. It was the last Gresley rebuilt Thompson Class A2/2 4-6-2 locomotive. Constructed at Doncaster Works as a semi-streamlined LNER Class P2 2-8-2 No. 2002 in October 1934, bearing Works No. 1796, it was rebuilt at Doncaster as Class A2/2 Pacific of orthodox design in August 1944, given the number 991 which subsequently was not carried, renumbered 502 during May 1946 and B.R. 60502 in June 1948. Allocated to York (50A) in November 1949 it was condemned while there during July 1961.
© *David P. Williams Colour Archive*

EASTERN TIMES • ISSUE 2

THE CLASS R 4-4-0s OF THE NORTH EASTERN RAILWAY

BY NICK DEACON

A dapper No. 1672 makes for an imposing portrait at an unrecorded date and location after receiving unlined black livery with the number being moved from the tender to the cabside, but before the red shaded lettering and numbers were replaced by Gill Sans characters. The loco dated from September 1907 and was one of the final batch of five produced that month. As related in the text the loco was for some years, up to 1914, credited with attaining the fastest recorded time on the Darlington to York section hauling the 12.20pm Newcastle to Bristol express over the 44¼ miles against a booked time of 43 minutes. In this view she retains

Designed by Wilson Worsdell and built from 1899 until 1907, this class was a response to the increase of main line express passenger loadings which the NER was facing from the latter part of the 1890s. With the GNR having introduced their first Atlantic class (the 'Klondykes') in 1898 to meet similar loading increases, the NER was under pressure to come up with their own solution. This duly appeared as a powerful 95 ton 2-inside cylinder 4-4-0 loco which incorporated many successful features already seen in Worsdell's earlier designs.

The first of the new 4-4-0s (No. 2011) appeared from Gateshead Works in August 1899 and was followed by a further twenty-nine of the class until May 1901. The locos were equipped with saturated boiler with an increased pressure of 200lb per sq. in. (later reduced to 180psi), 19 x 26 inch cylinders, a lengthened firebox of 7ft and a boiler diameter of 4ft 9in – the largest employed by the NER at that time. All the boilers and spares were fitted with Ramsbottom safety valves with the distinctive polished brass trumpet cover. The well-tried Stephenson motion was retained, but operated by newly-developed 8¾in

a 'windjabber' or capuchon to the rim of the chimney (although most had been removed by 1945), has Ross 'pop' safety valves, twin bell-shaped whistles mounted on a 'U' stand in front of the cab and the smokebox carries a Gresley-type anti-vacuum valve mounted behind the chimney. Receiving a 59A boiler in September 1946, in BR days as No. 62397 she gravitated at various times between Bridlington, Starbeck, York, Selby and Neville Hill sheds before a return to Bridlington (53D) in June 1956 which saw her lasting there until condemned in February 1957. *Photo: Neville Stead Collection (NS201646) © The Transport Treasury*

A classic pre-1914 view of the south end of York station featuring No. 2020 in original 'saturated' condition departing towards Holgate Junction with an Up passenger service comprising of four non-corridor clerestory coaches and one extra non-corridor elliptical roof vehicle added between the loco and the end brake vehicle. The loco was new to traffic during December 1899 and received a superheated boiler in May 1914 and much later, when allocated to Starbeck shed (50C) in 1936, it was the choice for an extensive rebuild as a D20/2 as mentioned in the text and seen in the accompanying post-conversion BR-era photograph. The train has just passed under the magnificent signal bridge with its multiplicity of arms indicating both 'Up' and 'Down' directions controlled by the 295-lever frame 'Locomotive Yard' signal box opened in 1909 which is out of sight to the left. The signal bridge also incorporated a footbridge passing over the running lines enabling access to the signal box. *Photo: Stephen Barrett Archive – SA Collection from www.Rail-Online.co.uk*

outside admission piston valves fitted below the cylinders. The class was also fitted with a deep 7ft length firebox giving a 139 sq.ft. heating surface with this placed ahead of the rear axle. With a good volume ashpan and dampers fore and aft it made for greater ease of firing when employing either a uniform fire-bed for maximum power or a tapered fire which only needed feeding from the back. All were dual-braked with Westinghouse cylinders for both the engine and train, and a vacuum ejector for alternative train braking.

With the combination of a simple, sturdy construction wedded to a lively, free-steaming, and reliable power base the class quickly earned their spurs and proved to be an instant hit with the operational and maintenance staff alike. Arguably, in this author's opinion, the class was amongst the best pre-grouping 4-4-0s ever built.

From her delivery, No. 2011 was allocated with two regular crews charged with working the taxing 455 mile, six-day per week diagram from Newcastle to Edinburgh and return (including the southbound 'Flying Scotsman' express), plus the balance of the job to Leeds which was operated by the second crew. This duty continued over the next two years almost without a break and when No. 2011 was 'shopped' for general overhaul in March 1903 it was found that she had clocked up an incredible 284,182 miles since her first appearance. The closest to this figure attained (on average) by other members of the class was around 163,000 so No. 2011's total was doubly astonishing – albeit in some quarters viewed with considerable scepticism! Dr. Tuplin in his *North Eastern Steam* (George Allen & Unwin, 1970) also notes that as the loco was based at Gateshead shed and therefore close to the works this proximity to maintenance excellence may have helped to underpin this remarkable record. A footnote to this story also records that on withdrawal from Selby shed in February 1951 (now as BR No. 62340) her total mileage was said to have reached 2,017,586 or an average of almost 388,000 per annum – not a bad return for a lifespan of 52 years! A few of the class were also allocated to the NBR Haymarket shed for its Newcastle workings and between 1904 until 1914 were the locos of choice for these duties.

A further batch of thirty of the class appeared from Gateshead from September 1906 until September 1907 bringing the class total up to sixty – a number which remained intact until the withdrawal of ex-NER No. 1147 in January 1943. One of the last batch of five built in September 1907 (No. 1235) appeared with an extended smokebox containing a Sisterton superheater which was trialled with a dynamometer car in March 1908. A further trial with a larger superheater occurred during April/May 1909 but clearly the results were not encouraging and the equipment was removed soon after but leaving the loco with its extended smokebox until September 1911. The very last of the class to be built in September 1907, No. 1672 (BR No. 62397), for some years until at least 1914, held the accolade of attaining while still non-superheated

On the Ripon–Harrogate line at a location known as Monkton Moor near the village of Bishop's Monkton, No. 62392 of Starbeck shed (50D) assists A3 Pacific No. 60074 *Harvester* of Neville Hill shed (50B) with a southbound Leeds express. The shot probably dates to 1949 when the D20 had been renumbered during the previous November but prior to its move to Selby in June 1951. However its tender still bears the legend of its previous ownership which was a commonly seen 'transition' pairing. Starbeck's D20s were often used as passenger train pilots, especially during holiday seasons which generated heavy excursion traffic. Starting life in September 1907 as NER No. 1235 it (like No. 1672 also illustrated) was one of the final batch of five produced that month which were the last additions to the class. It was superheated in November 1913, reboilered with a 59A in September 1945 and lasted at Selby until withdrawn during May 1954. The location with the bridge in the background carrying a minor road into the village of Bishop's Monkton was apparently favoured by photographers as the gradient here could usually be relied on to produce a good exhaust although not on this occasion!
Photo: Neville Stead Collection (NS201628) © Transport Treasury

the fastest recorded time on the Darlington to York section hauling the 12.20pm Newcastle to Bristol express with 165tons on in 39min. 34 seconds over the 44¼ miles against a booked time of 43 minutes. Times of 41 and 42 minutes had already been recorded with this service (advertised at the time as 'The Fastest Train in the British Empire') but as far as is known No. 1672's effort stood until eclipsed by her LNER successors.

All bar two of the class were superheated with lengthened smokeboxes from 1912 onwards during Vincent Raven's time as the NER's CME with the balance dealt with by the LNER in 1925 and 1929. Most of the first superheated boilers included Schmidt superheaters, but seven were built with Robinson variants instead. The Schmidt superheaters remained standard until 1930, when a switch was made to LNER-standard Robinson superheaters.

The opportunity was also taken to fit mechanical lubricators for the cylinders at the same time that superheaters were fitted. The fitting of superheaters required the smokebox be extended by about 1ft in length and so an extra section of frame was welded into the front curve of the frame section to correspond with the extended smokebox.

From 1916 onwards the Ramsbottom safety valves were replaced by Ross 'Pop' safety valves and this process was continued by the LNER until 1936 when the last loco to be so treated (No. 2012) was repaired and received a new boiler at the same time. The attractive NER chimney capuchons (or 'windjabbers' as they were more familiarly known) were retained until the mid-1930's but removed by 1945, except for No. 2110 which retained hers until withdrawal in March 1951.

On 16th August 1947 the north end of York station and the Royal Station Hotel are the backdrop to No. 2342 heading a stopper service possibly to Scarborough. Originally NER No. 2013, the loco was the third to appear in September 1899 as part of the first batch of ten to enter service between August and December of that year. It was superheated in October 1912, received a 59A boiler in March 1937 and became LNER No. 2342 in September 1946. It was also one of the class to receive a sight screen positioned between the two cabside windows and at this date still retains the brass beading to the splasher rims. After a period in store at Scarborough shed it moved to Selby shed (50C) in October 1950 to become one of the throng of the class allocated there but was one of the casualties of the 1951 cull of the class at the shed which saw it departing for scrapping at Darlington during March of that year.
Photo: Neville Stead Collection (NS201554) © Transport Treasury

At North Road, Darlington on 4th July 1948, No. 62340 demonstrates her 'ex-Works condition' and reveals the first and, it has to be said, woefully unimaginative, BR ownership 'logo' (if one can call it that) applied to the tender. In slight mitigation it has to be said that the newly formed merger possessed no distinguishing emblem at all until the 'Cycling Lion' started to appear from 1950. The loco also awaits the application of her Selby (50C) shedplate where she was based at this time. As NER No. 2011 she was the first of the class to enter traffic in August 1899 based at Leeds Neville Hill shed and her first duties were with two regular crews on a tough Newcastle to Edinburgh express and back and thence a return to Leeds with the second crew. She was also one of the star performers on 'The Fastest Train in the British Empire'. Unfortunately the July 1948 visit to the Works was probably her last prior to the 'final' visit as she was condemned from Selby during February 1951 – probably being surplus to requirements as the shed had no less than thirteen of the class on its books at this time. Condemned in the same month at Selby were No. 62348 and another four succumbed in March.
Photo: Neville Stead Collection (NS201551) © Transport Treasury

Under LNER management (and now reclassified as D20), the class changed from NER to LNER green livery and retained the semi-circular brass beading around the splashers but from 1928 as each needed repainting a change to black with a red lining was adopted until the latter was gradually dropped from 1941. From the post-Great War/early Grouping period principal express services were increasingly worked by the Raven Class Z (LNER C7) Atlantics which had been introduced from 1911, but the 4-4-0s still had main line duties based at the larger sheds. However, from around 1927 the class began to migrate to more secondary duties largely resulting from the introduction of the new Gresley D49 4-4-0's from that year and also because of the cascading of the Raven Atlantic locos onto former D20 duties – this in turn caused by increasing numbers of LNER Pacifics appearing during the mid-1920s. These secondary duties typically included services to Doncaster, Hull, Leeds, Scarborough, and York; Leeds to Northallerton and West Hartlepool and also Newcastle to Carlisle. The class were also regularly used on the services from Leeds to Harrogate and Northallerton with this route becoming something of a preserve of the class with at one time over half of the class allocated to the sheds which operated the route – these being Neville Hill, Starbeck, Stockton, West Hartlepool and York. Another indication of the reduced status of the class came in 1937 with an instruction to remove the tender water scoops, but this was hardly seen as an imperative at sheds as some were still being removed as late as 1945.

Despite the class getting somewhat long in the tooth by the 1930s, the LNER decided to prolong the life of these very capable locos with some modifications undertaken from 1934-35. The first of these involved the introduction of the boiler 59A design which incorporated a single plate barrel instead of the original three ring design plus the addition of sixteen tubes which enlarged the heating surface by 87 sq. ft. LNER-pattern large 'dished' smokebox doors were sometimes fitted as boilers were changed. These changes were followed from 1936 with piston valve modifications including a repositioning above a new set of cylinders, a valve enlargement from 8¾in. to 10in. and an increase of the steam and travel laps. The loco selected as the test bed, No. 2020 (BR No. 62349), was also converted to left-hand drive, lost its Westinghouse braking and, uniquely for the class, also acquired raised footplating, separate splashers, a new cab with sighting screens and revised tender body shape with coping stepped out along all sides. Although Gresley was the LNER CME at the time, the initiative for the 1936 modifications had been instigated independently by his subordinate, Edward Thompson, whilst he was Mechanical Engineer at Darlington. Possibly because he had not been consulted prior to the work commencing and that it had been given advance coverage in the technical press, Gresley admonished Thompson for his presumption and vetoed the changes to other members of the class. There the matter rested until Thompson

The class were no strangers north of the border and in the years leading up to 1914 Haymarket shed, Edinburgh was apt to rely on its small stud of the class for its rostered services to Newcastle. However, by the 1940s/50s the class were seen only on branch services feeding into the Waverley line such as those from Berwick to St Boswells, Hexham to Riccarton/Hawick and Reston to St. Boswells via Duns. At the south end of Hawick station No. 62387, an April 1907 build and superheated in October 1917, heads a service for Newcastle via Riccarton and Hexham. It had received a 59A boiler in May 1948 and an LNER-pattern large 'dish' shaped smokebox door. Helping to narrow the photographic time frame down a little, from October 1953 until July 1955 the loco was the sole representative of the class based at Heaton shed (52B) and known to have been rostered on services to Hawick until a two-year move further south to Selby occurred. A last move to Tweedmouth and Alnmouth saw her employed on two society tours during June 1957 prior to withdrawal during the first week of September. There was possibly some thought of retaining her for a while at least as an observer at Darlington Works in November saw her partially greased up as if this was the intention. Alas, events proved otherwise and it was not to be.
Photo: Neville Stead Collection (NS206069) © Transport Treasury

At the north end of Alnmouth station sometime during the 1950s No. 62349, then allocated to Tweedmouth shed (52D) or its 'sub' at Alnmouth, waits to depart with a main line 'stopper' to Tweedmouth or a branch service to Alnwick. New to traffic during December 1899 as No. 2020, in October 1936 the loco had emerged from Darlington Works as the only member of the class to receive a combination of modifications with the new LNER 59A boiler – these being new cylinders with a piston valve enlargement from 8¾in. to 10in., and with the latter now fitted above the cylinders and not below, an increase of the steam and travel laps, left hand drive, raised footplating above the driving wheels, separate splashers, a new cab with sighting screens (although this appears to be missing in the photograph) and a revised tender body shape. As a Starbeck engine prior to the conversion it was eventually returned home to its regular duties which included haulage of the 11.15am 'Yorkshire Pullman' between Harrogate and Doncaster. It remained in its converted form and was withdrawn from Selby shed (50C) in February 1956. *Photo: Neville Stead Collection (NS201642) © Transport Treasury*

became CME in 1941 but thereafter only three more of the class were actually modified but without changes to the footplating, splashers, cab and tender. The locos affected were No. 592 (BR No. 62371) in October 1942, No. 2101 (BR No. 62360) in December 1942, and No. 712 (BR No. 62375) in October 1948. The rebuilt locos were re-classified as D20/2 and the un-rebuilt original locos became D20/1. The unique No. 62349 lasted until February 1956 when it was withdrawn from Selby shed. Still complete, she was observed at Gateshead shed during May but by the end of the month she had been moved to Darlington for scrapping.

Prior to Nationalisation in 1948 the class had endured gruelling years of hard work resulting from heavy wartime demands and many had been allowed to get into poor condition which had seen nine of the class making their last visits to the 'melts' at Darlington following the unfortunate No. 1147 which was the first to go during January 1943.

Despite being seen less frequently on principal main line duties the class continued to play an important role on summer excursion traffic (which had built up again since the end of WWII), piloting LNER D49 4-4-0s, V2 2-6-2s and Pacifics on main line services and many secondary passenger duties and goods work. The class was particularly associated with summer traffic rostered from Selby shed to resorts such as Bridlington, Filey and

Two for the price of one at Scarborough on 4th September 1954. Double-headed by Nos. 62378 and 62374 (the latter with a cabside window sight screen) in workaday grime wait to depart from Londesborough Road station with a heavy returning excursion service which they will head as far as York or Selby or even further on to Leeds. Both locos were long-term Selby shed (50C) inmates since at least 1948 and both were destined to be withdrawn from there with No. 62374 succumbing during the following October after the end of the summer season, and No. 62378 hanging on until November 1956. Londesborough Road station was opened on 8th June 1908 as 'Washbeck Excursion Station' to relieve holiday season pressure experienced at the main terminus. It was upgraded to a public station, renamed 'Londesborough Road' from 1st June 1933 and appeared in the following month's Bradshaw. It closed as from 24th August 1963. The train is seen at the 300 yard through platform which most excursion traffic used. To the right of the train through lines continue to the main terminus which is ½ mile further on. Photo: Neville Stead Collection (NS201599) © Transport Treasury

Scarborough and also tightly timed local stopping passenger services to Doncaster, Harrogate, Hull, and Leeds and from time to time were even recorded on goods and passenger services over the now sadly diminished Hull & Barnsley line. On Vesting Day 1948 Starbeck shed had the biggest number of the class on its books with ten, closely followed by Selby with nine and Alnmouth with seven. The balance was scattered around other sheds in the NE area with Hull Botanic Gardens claiming most with five.

From May 1949 until June 1950 ten of the class were given new 3,900 gallon and 6ton 5cwt coal capacity straight-sided tenders mounted on the existing frames and wheels. The new (and scoop-less) tenders were shifted around a variety of the class over the years and when seen wedded to their NER steeds of decidedly Edwardian appearance their 'modern' look tended to look somewhat incongruous.

By 1957, the last of the class (BR No's 62375/81/3/7/95/6) had gravitated to Tweedmouth shed and its 'sub shed' since 1950 at Alnmouth and were operating the local Alnwick services into Newcastle plus the Alnmouth branch line. No's 62375 and 62383 departed for the final trip to Darlington in May and April '57 respectively – despite both reported as being in good external condition. In June No. 62387 enjoyed some brief limelight and been nicely 'bulled up' to head a couple of society tours: firstly on the 2nd for 'The Yorkshireman' Branch Line Society tour from York, and on the 23rd for the first leg from Leeds City to York on

Another long-term Selby allocated D20, this time No. 62386, departs from Hull Paragon station with a stopping service to Selby and Leeds or York. The loco was one of Selby's allocation which were regularly used on services within an area bounded by Doncaster, Hull, Leeds, Harrogate and York plus excursions to the coastal resorts of Bridlington, Filey and Scarborough and also when called on to render pilot assistance on heavily loaded cross-country services. As NER No. 1207 it was new to traffic during August 1907, was superheated during April 1913 and was one of the locos which regularly attained fast timings on the Darlington to York section whilst employed on the 12.20pm Newcastle to Bristol express referred to in the text. The loco was reboilered to a 59A type in October 1943, received a cabside window sight screen after 1947 and remained at Selby until withdrawn during October 1956.
Photo: Neville Stead Collection (NS201614) © Transport Treasury

the RCTS Yorkshire Coast Rail Tour. It also headed the return leg from York. On both occasions despite being a tad run down the loco ran well and kept time. In August a leak developed in the tender and it was observed travelling through Heaton on the 16th en route to Darlington Works for attention. However, soon after arrival at Darlington it was condemned during the first week of September. This was followed by the last three of the class (BR Nos. 62381/95/96) succumbing during November and December 1957 with the very last to go, No. 62396, having totted up a lifespan of 50 years 2 months. Regrettably (and shamefully), none of the class were preserved.

A modest swan-song of sorts was accorded to No. 62395 when, on arrival at Darlington for scrapping, she was repainted on one side only and given the new BR heraldic emblem for an official publicity photograph. It would be interesting to know what the circumstances were in authorising this – perhaps as none of the class were to be preserved it was a local initiative to record the demise of this much-loved class.

This short account of the class is largely based on facts derived from the RCTS Locomotives of the LNER Part 3C, published in 1981, various editions of The Railway Observer, the 'BRDatabase' website and with additional material gleaned from North Eastern Steam by W. A. Tuplin published by George Allen & Unwin in 1970 and North Eastern Locomotive Sheds by Ken Hoole published by David & Charles in 1972.

EASTERN TIMES • ISSUE 2

TALES OF A B17 SPOTTER ON THE STOUR VALLEY LINE
BY JOHN BUTCHER (B17SLT)

Long Melford Junction signal box and station. *Photo: Roy Edgar Vincent (REV82A-3-6)* © The Transport Treasury

The following article about the Class B17 4-6-0 locomotives is based on my own spotting observations and experiences on The Stour Valley Line linking Colchester and Cambridge, it is factual as I was living in Long Melford at the time. I have written it because I think it affects the history of the B17 class.

In simple terms, I am trying to find any other witnesses who can corroborate that B17s were running in service after official withdrawal dates. The withdrawal date of the last locomotive is listed as August 1960 for No. 61668 *Bradford City*.

My write up mentions three witnesses other than myself but I personally do not know how to try and track them down. Anyone who travelled on the last service between Long Melford Junction and Bury St. Edmunds in April 1961 (62 years ago) were certainly witnesses to the event. Also train drivers and train spotters at Colchester and Cambridge stations during the 1960/61 period may remember viewing B17s in steam. Additionally similar travellers on the Colchester to Clacton branch could have observed a B17 'Sandie' near the Paxman Works at St. Botolph's (Colchester Town) station.

THE EXTENDED LIFE OF CLASS B17s AS WITNESSED AT LONG MELFORD JUNCTION

1. THE EARLY YEARS

My Parents lived in Long Melford, Suffolk. Shortly before I was due to be born in early 1947 there were heavy snowstorms that blocked the road, with 7 foot snowdrifts on the route to the maternity hospital in Bury St. Edmunds. Consequently, my Mother was taken the three miles to Sudbury where I was born.

There was then a branch line between Long Melford Junction and Bury and prior to the Beeching Report cuts, was scheduled for closure on 10th April 1961. For at least the three years prior to this date, dieselisation was actively taking place on the Eastern Region of British Railways including the Stour Valley Line and its branch services.

Under the BR Modernisation Plan (1955), the Region chose diesels with diesel-electric transmissions; it was the first region to begin replacing whole classes of steam (e.g. Class B17) with diesel locos on the main line plus adding Diesel Multiple Units (DMUs) on branch lines. Electric Multiple Units (EMUs) and locos came later on the main lines. Incongruously as described later, dieselisation gave some steam locos a longer life than planned.

2. B17 SPOTTING

However, my train spotting hobby did begin at Long Melford near Liston Lane level crossing at the end of my road, this being on the Stour Valley Line. This momentous event happened at approximately the tender age of 10 (1957/58). I did not really understand much about the steam engines that I was seeing but I am now sure that I saw the following eight locos on general everyday services that consisted of only two or three carriages – Nos. 61639 *Norwich City*, 61650 *Grimsby Town*, 61651 *Derby County*, 61654 *Sunderland*, 61662 *Manchester United*, 61666 *Nottingham Forest*, 61667 *Bradford* and 61668 *Bradford City*. They were all allocated at the time to Colchester (30E) steam shed before its closure in November 1959.

The crossing keeper suggested that I visit Long Melford station to get info and view timetables. One day after months of indecision, I got on my bike and headed there as I had

Class B17 4-6-0 No. 61666 *Nottingham Forest* at Long Melford in July 1959.
Photo: Courtesy John Butcher

Class B17 No. 61653 *Huddersfield Town* approaching Long Melford in late 50s with a Clacton Excursion.
Photo: Courtesy John Butcher

begun to notice boring DMUs on occasional services as the dieselisation policy was being implemented.

I then discovered that the station was a junction with a branch line! At some point in 1958 on Long Melford station I met a lad of my age (John Boulden) whose family had recently moved from London as part of an overspill scheme. He was clued-up on railway locomotives and train spotting, especially in the London area.

By chance he was meeting the 3.36pm to Cambridge as it had a pile of Evening Standards on board for him to deliver. Consequently, I was made aware what a Class B17 looked like as there was one at the head of a train. He confirmed what it was with an Ian Allan book containing B17 pictures and technical detail when we met at the station on another occasion.

So as witnessed by myself, B17 '*Sandies*' were hauling this aforementioned Colchester-Cambridge daily service whenever I was spotting (1958 to 1961). A DMU has never run on this particular service; it remained loco hauled up to the line's closure in 1967.

I can recall that one day I was talking to the driver of 'Football' loco No. 61667 *Bradford* as it filled with water before heading towards Sudbury/Colchester. He said that it was being withdrawn from service and sent for scrapping the next day. I asked what was wrong with it and he stated "nothing!". Therefore, this was June 1958, believing the record for this B17, as it seems correct. I also remember viewing No. 61668 *Bradford City* on the same duty about a week before because the similarity of the loco names did affect my grey matter.

NOTE: Remember that the name '*Sandie*' is an affectionate nickname for the B17 Sandringham class of 4-6-0 express steam locomotives. Also '*Footy*' is used for the later B17s named after football clubs.

3. ENTHUSIAST WITH TRAIN SPOTTING CREDENTIALS

By the end of 1958, I was very aware and enthusiastic about train spotting as Mr. Boulden plus his father kindly escorted me to their hometown of London where we bashed all the main loco sheds and visited main line stations plus Clapham Junction. The steam sheds included 1A Willesden, 1B Camden, 1C Neasden, 14B Kentish Town, 70A Nine Elms, 81A Old Oak Common plus 73C Hither Green Diesel Depot. The stations included all the large main line terminals plus the infamous LMR Broad Street station high above Liverpool St. Station. I must credit them with stimulating my train spotting and engineering interest that eventually led to me becoming an electrical engineer. Consequently, I appreciated all the locos on view at Long Melford Junction and others viewed in London. Sometime in early 1959, I found Class J15 No. 65361 in a station siding at Long Melford station with one connecting rod from the wheels placed in the cabin, indicating that it had failed. Though this did not look good for it at the time, the loco has the honour of being the last J15 to be withdrawn when Stratford (30A) shed was closed in September 1962.

For the record, at Liverpool Street station in August 1958 on the return leg of our trip to London, we spotted the one Clan class loco from the Scottish Division that had been loaned to the Great Eastern; it only managed to stay on the Great Eastern main line for one month before it was returned. It was No. 72009 *Clan Stewart*. Unlike the similar Britannia Class (e.g. 70000), it did not like Brentwood Bank. We saw it in action on the turntable.

At St. Pancras on the same trip, we had seen 4-car suburban DMU sets for commuters, normally in an 8-car formation. One Sunday in 1959 I was walking my dog in meadows near my local Long Melford, Liston Crossing so imagine my surprise when three of these sets passed by, a 12-car train. I had come across a Bedford-Clacton excursion. A very unusual visitor to the Stour Valley Line and Clacton although running weekly that Summer.

To further prove my train spotting credentials in this era, I know of just one photo taken of me viewing J15 No. 65450 in September 1959 at Long Melford station (below). I found it on the internet about two years ago and remember the photographer making me aware that I was in a photo, I asked how I could obtain a copy but he walked off!

The J15 was waiting with two coaches for a later Bury branch service. So this one photo also proves that steam was again present on the Bury branch after the 'official' July 1959 end of steam fanfare. The diesel loco replacements were proving to have poor availability.

4. STOUR VALLEY LINE AND BRANCH SERVICES

4.1 History

In the early 1950s, the mainstay for passenger train power on the Stour Valley Line of the Great Eastern were classes E4 2-4-0, J15 0-6-0, K3 2-6-0, D16 4-4-0 and Midland Ivatt Class 2MT 2-6-0 'Flying Pigs'. Cambridge shed (31A) and Colchester (30E) had at least five of the latter between them. In the later 50s, class B17 4-6-0s, displaced from the GE main lines, began to take over these passenger workings prior to there being sufficient DMU availability.

Occasionally trains could produce 4-6-0s of classes B17, B12 or B1, especially on excursions or specials, but normally these 4-6-0s ran on the main line until diesel locos began to take over their duties. In the late 1950s the short Long Melford to Bury branch had passenger trains hauled by Class E4 2-4-0 or J15 0-6-0 steam locos, but it often saw Class F6 2-4-2T or C12 4-4-2T locos in the early 1950s. There was always water at Long Melford if these small locos needed it.

At the time that my spotting interest started (1957-58), the use of J15s, 2MTs and E4s on the through trains on the main Cambridge-Colchester Stour Valley Line was ceasing for regular passenger work. Some DMUs were beginning to appear and B17s had become available at 30E shed. Maybe drivers then ensured that B17s got priority use over the smaller locos when carriage stock was needed. On the Stour Valley Line, B17s generally ran with just two or three carriages so only needed to be lightly fired.

J15 No. 65450 has just arrived from Bury on 1st September 1959, the 12 year old me is on the Long Melford up platform. Photo: Courtesy John Nunn, Long Melford Heritage Centre.

Holden 'Claud Hamilton' Class D16/3 4-4-0 No. 62618 arrives at Long Melford with the 2.25pm service from Marks Tey to Cambridge on 21st July 1956. By this date it was displaced from 31A Royal Duties by B17 61671 *Royal Sovereign*.
Photo: Leslie Freeman (LRF2355) @ The Transport Treasury

Dieselisation slowly began to bite. Initially the small Waggon & Maschinenbau 4-wheel railbus was used for the occasional off-peak Stour Valley service instead of a steam loco and carriages; however soon these railbuses were relegated to the minor branch lines, such as the nearby Colne Valley and Saffron Walden branches. Cambridge (31A) had an allocation of five. Very gradually throughout 1959, the first-generation DMUs of Wickham (Class 109), Craven (Class 105), Metropolitan-Cammell (Class 101) and Derby lightweight manufacture took over from the B17s on local Colchester (30E) turns. The DMUs initially had teething problems.

Freight was still seeing a regular Class J17/J19 0-6-0, and even occasionally the more powerful J20 0-6-0 locos, appearing on the daily turn to Sudbury Goods Yard. The latter locos has a boiler identical to the Class B12 4-6-0 express locos. There are even pictures showing J17s and J19s on Stour Valley passenger workings due to motive power shortages. By the time the daily goods reached Cambridge, it often had up to a hundred loaded goods wagons, something I witnessed twice. The sugar beet season really filled the wagons. In addition to the Sudbury

Class J19 0-6-0 No. 64656 shunting in Long Melford Goods Yard.
Photo: Dr. Ian C. Allen (ICA E3650) © The Transport Treasury

Goods, a J20 0-6-0 or WD 8F 2-8-0 March (31B) freight working ran over the Bury-Long Melford branch every morning via Sudbury Up to Marks Tey and beyond, with an evening return service. Use of the Bury and Stour Valley routes for goods/passenger through trains eliminated train reversal to some destinations, for example Harwich and Clacton.

4.2 Bury Branch Workings

During my early visits to Long Melford Junction, the local Long Melford branch line to Bury St. Edmunds normally still had a daily Class J15 from 31A on the two or three carriage service. There was a fanfare on the occasion of the last "official" steam service to Bury in July 1959 and it was J15 hauled (tender first), this proved to be wrong on several occasions.

Ipswich (32B) lost its last five B17s in 1959. The shed was just receiving an allocation of North British (NBL) BR Class 21 Type 2s (D61xx). So from then onwards, I observed these brand new diesel-electric locos hauling the passenger set for many months. However I noticed it was still hauled by class J15 locos intermittently; there were certainly problems with the reliability of the NBL locos and maybe replacement DMUs were in short supply. Whenever possible, it seemed that passengers still liked the comfort of carriages.

The choice of the NBL class around this time for the branch services proved to be very significant, a factor that affected other workings at Long Melford. Here I must allude to an event that I believe is related to the phenomenon of B17s coming back from the dead; refer to Section 4.5 for more facts.

The NBL diesel locos running the trains to Bury were getting a reputation for poor daily availability with numerous oil leaks, engine failures and even fires. The important event is namely that BR finally decided to transfer all of the NBL locos back to Scotland and Eastfield

Class J15 0-6-0 No. 65477 approaching Long Melford.
Photo: Dr. Ian C. Allen (ICA E4007) © The Transport Treasury

Class B17 No. 61668 Bradford City at Long Melford in May 1959.
Photo: Norris Forrest (NF005/10) @ The Transport Treasury

Depot (65A), a sensible choice for warranty reasons as this would be close to where they were built. Significantly, this event took place gradually during 1960. This resulted in motive power shortages in the Eastern Region as there were not enough diesel locomotives available.

4.3 Regular B17 Services

However on the Stour Valley Line in 1958 and later times, there were still two steam-hauled passenger services daily (except Sunday) with each comprising four or five coaches hauled by a B17. The Down (2L02) Colchester-Cambridge (ex-Long Melford, dep. 3.38pm) and the later Up (2F05) Cambridge-Colchester (ex-Long Melford, dep. 6.30pm). There was a different steam loco for each train. The Down loco always faced the correct way at Long Melford because there was a triangle near St Botolph's terminus station (now named Colchester Town) and service facilities near the Paxman Works; this was only two miles down the Clacton branch from Colchester main line station, so at least one B17 spent the night here..

In the Summer, there was also a timetabled Saturdays only (SO) 1E59 Leicester to Clacton excursion, passing Long Melford at 11.45am and a 1M59 Clacton to Leicester passing Long Melford at 3.09pm. The services ran from 10th June until 15th September and generally comprised around 10 Carriages.

I always saw a B17 'Sandie' on these four services and no other class. I was just too late to see Class K3 or D16 workings on the excursion services. It is significant that these duties were by 31A/31B crews, so the train personnel were based in the Cambridge area.

The two particular weekday daily services customarily used carriage sets and this continuing preference proved to be a very significant factor in prolonging the working life of several Class B17 locos. Sufficient DMUs were not available so not considered for the two daily services described here. I assume that there were no spare Brush BR Class 31 Type 2 (D55xx series) locos as none appeared on any Stour Valley or Bury branch passenger service, based on my observations.

4.4 Sandie Ghosts

On several occasions when we (John Boulden and myself) viewed one of the daily B17 services together over approximately the next 30 months onwards from mid-1959, he looked at his books and told me that "this loco is supposedly withdrawn! This engine is a ghost!" They certainly were NOT scrapped at the time despite records to the contrary! Over the following months, we witnessed several of them running in a serviceable state hauling trains but after documented disposal dates. This was especially true after the last B17 No. 61668 *Bradford City* was supposedly withdrawn. Every Weekday from then on after 22nd August 1960, we were seeing ghost B17s! This continued for more than seven months after the August date.

I did not have a spotting book at this time to record/remember the numbers but I do recall that most of the locos had football club names (apart from *Clumber* at least). So personally, I believe that I have not seen many of the class that were constructed before 1936-built No. 61648 *Arsenal*, one the first of the 24 *'Footy'* locos.

So any day (except Sundays) when convenient, I could witness at least two B17s at the Liston Lane level crossing near my home. Unfortunately, I was not into cameras as I had little money so I took no photos. I had not yet started earning from my paper round.

4.5 The Class 21 Factor

The transfer of NBL Class 21 diesels back to their northern roots caused a shortage of locos in East Anglia. Presumably, the Class 31s were rostered elsewhere, surely covering for the NBL Type 2s sent to Scotland and additionally the loss

Class B17/6 4-6-0 No. 61623 *Lambton Castle* at Long Melford Junction. The branch to Bury St. Edmunds is on the right.
Photo: Dr. Ian C. Allen (ICA E795) @ The Transport Treasury

I copped NBL Class 21 Type 2 Diesel-Electric D6122 at Long Melford station. Here it is seen in the mid-1960s passing through Princes Street Gardens, Edinburgh after transfer to Scotland.
Photo: George C. Bett (GCB609) © The Transport Treasury

of thirteen 'Sandies' from 31A during 1959/early 1960, cut up far too quickly by Doncaster Works. Consequently as a result of almost no Class 31s available to haul Stour Valley passenger services (certainly there were none allocated to Cambridge), resulted in the need for some steam locos to continue running or return to service due to these motive power shortages.

5. COLLECTING OUR THOUGHTS

The two B17 daily steam-hauled services through Long Melford continued to run without any interruption during 1960. This continued for months after the last B17 No. 61668 *Bradford City* was officially withdrawn in August of that year. A last bastion for a B17 service borne out of necessity. Whenever I had the time to view at Long Melford Junction or Liston Crossing, a B17 was hauling the service.

6. CLASS E4 62785

The return to service phenomenon cannot be better illustrated than quoting my first spot of a Class E4 2-4-0, namely No. 62785 in February 1961. I witnessed her arriving as the daily loco on the Bury service. She was officially withdrawn in November 1959 from Cambridge shed (31A). At that time, she was the last remaining of 100 built and the last 2-4-0 main line steam loco operating on British Railways. I found out from the crew that she had returned to this service in late 1960. What a choice of steam loco to return to service albeit unofficially.

So the last steam service on the Bury branch was not the fanfare event in July 1959; it transpired to be in March 1961, this time without a fanfare! This running of No. 62785 has never been acknowledged officially, but I definitely witnessed it. She disappeared in early March 1961 when a DMU became available. I decided to visit Cambridge shed and investigate further. This Branch Line was due to close on 10th April 1961. More later about that special day in section 9.

The last steam loco on the Long Melford to Lavenham line was Class B1 4-6-0 No. 61287 that was used to haul the rail removal train in May 1962. The work took just a few days.

7. 31A SHED BASH

So on 25th March 1961 I visited Cambridge (31A) for a day of train spotting. I set out on the first DMU of the morning to get in a full day on Cambridge station. I had B17s and

Holden Class E4 2-4-0 No. 62785 at Long Melford Up platform.
Photo: Dr. Ian C Allen (ICA E527) © The Transport Treasury

Class B1 4-6-0 No. 61287 at the head of the Permanent Way train removing track between Long Melford and Lavenham.
Photo: Dr. Ian C. Allen (ICA E1141) © The Transport Treasury

the E4 in my mind, amongst other locos generally. In those days, the BR Class 23 ('Baby Deltics', D59xx numbers) were running the services to Kings Cross. I also witnessed Class D16 4-4-0 'Claud Hamilton' No. 62613 immaculately clean on a Kettering service and a Patriot, with a Nuneaton (2B) shed plate, on a goods train that must have run over part of the old Cambridge-Oxford Line that is now being effectively reinstated as the East-West Line. The Class D16 was my one and only cop of the class. It was the last in service when supposedly withdrawn in September 1960 from 31B. Was this another Ghost Loco?

I decided to bash Cambridge shed (31A) for the first time ever. I remember that there was another Patriot present, out of steam, so two ex-LMS Patriots at Cambridge on the same day must be a bit unusual. Also there were no B17s on shed which indicated to me that 31B March shed must have been maintaining the remaining B17s for the months I had been viewing at Long Melford. A very enthusiastic Doncaster Works had decimated the 31A B17 allocation of 13 locomotives in 1959/January 1960.

Lo and behold I found the E4 No. 62785 at the very back of the shed in its 'preservation position', it was scheduled for the National Collection. I have been told since by a B17SLT member who witnessed it, that it was the last steam loco to leave 31A and under its own steam. It is at Bressingham Steam Museum with its original GER number 490 on it, carried when it had the class designation T26. It was refurbished and painted in its original GE blue livery at Stratford Works before it was put on display. I must go and view her again sometime as over 60 years have passed since I last saw her, and I like blue!

8. A CELEBRITY SPOTTER

I digress a bit from the B17 main topic of interest now. On the Cambridge station platforms that day were other spotters and I recognised one of them, although I never knew his name. I had seen him on the local BBC TV news posing in front of 9F 2-10-0 No. 92220 *Evening Star*, the last BR steam loco built and completed in 1960. It was supposedly the last number he needed to complete the spotting of all BR steam locos, a full house. He told me that this was not quite true and admitted to never having seen B17 No. 61620 *Clumber* but he knew that it had been withdrawn. When I told him I lived in Long Melford, he said that he would be visiting there for the last service to Bury St. Edmunds on 10th April 1961, which was two weeks later. I told him that he would see a B17-hauled train, but he did not seem to believe me. Perhaps he knew that they were all officially withdrawn. Talk about waving a red rag at a bull – I now had a mission.

9. B17 'MISSION CLUMBER'

So to my mission. Later that same day, the regular B17 appeared heading the Colchester service and I spoke to the engine crew at Cambridge station and told them the story of the TV spotter and mentioned to them that they had two weeks to get No. 61620 *Clumber* up and running if they could. They said nothing and I had no idea if it had been cut-up or not. The TV spotter had disappeared somewhere. The train had come in from the March direction so I assume that 31B was servicing the stock and the remaining B17s.

10th April 1961 arrived and there was a boring Derby Lightweight DMU provided for the last passenger train to

Bury St. Edmunds. Before it departed, the 3.38pm loco-hauled passenger train to Cambridge had to arrive from the Sudbury direction. My celebrity friend train spotter was there as he spotted me on the footbridge looking towards Colchester; I am aged 14 by this time.

I said "you see the smoke in the distance? Well that is *Clumber*!". He said nothing but I was just making a statement that I wanted to come true, but not knowing anything for sure! Unbelievably No. 61620 *Clumber* arrived and I left him to revel in an unbelievable dream come true, his last steam spot, a B17 and at my home station. According to records, *Clumber* had been withdrawn on 25th January 1960. This obviously was not true, nor the disposal date of 31st January 1960.

The TV cameras were not there to catch *Clumber* running that day, I was one witness who did, but there were others.

10. CONCLUSIONS

10.1 Clumber and beyond

I stopped visiting Long Melford station from that day onwards as I felt that everything had changed, it was no longer a junction. I do not know the end date for the B17s on the Stour Valley line, the driver of *Clumber* on that memorable April day reckoned that it was steaming well and predicted that "it will be running for weeks more". I know that these Stour Valley steam-hauled services were the last regular B17 turns and they gave the class an extended life, thanks mainly to the reliability problems of North British diesels.

10.2 False auditing and dates

In 1959, Doncaster Works performed the scrapping of all of the B17s, maybe because Stratford Works said no to killing off a Great Eastern product; nine pre-1936 and four 'Footy' locos were culled from the Cambridge 31A allocation and the last four of the Ipswich 32B allocation met the same fate. 31A lost its last three B17s in January 1960.

However records for 1960 show that thirteen of the final fifteen B17s (all B17/6s) are listed as being scrapped or cut-up at Stratford Works, including No. 61608 *Gunton*, the oldest survivor. The 'disposal date' information for most of these withdrawals does not reflect what I saw on the ground at Long Melford. The department responsible seems to have had foresight and appear to have cooked the books regarding cut dates to keep management happy. In reality, to address a short-term motive power shortage

Class J17 0-6-0 No 65564 on a freight working takes the Bury line at Long Melford Junction, however it is only using the branch as part of a regular shunting movement into a siding. It will depart for Cambridge. *Photo: Dr. Ian C. Allen (ICA E3792) © The Transport Treasury*

some locos continued running in service; I suspect they were based at 31B March as it was reasonably close to the Stour Valley Line.

10.3 Post B17 Services

On a few occasions in later years I returned to Long Melford and the Stour Valley Line to witness what looked like the same coaching stock being pulled by Paxman BR Class 15 Type 1s (D82xx series) and possibly these remained on the same services until closure in March 1967. I was away at University from 1964 so I missed this sad event. I was now a full-time student with no time to be a railway enthusiast. I have since made a comeback!

10.4 Other witnesses

On 26th September 2021 I was visiting the Epping and Ongar Railway for a diesel weekend and got talking to an old enthusiast with an expensive camera. I mentioned Long Melford and he told me that he travelled on that last branch service to Bury St. Edmunds. He also remembered "a well turned out steam loco" at Long Melford but he preferred diesels! Before I could ask the question, he legged it so that he could take his next photo. He didn't mention that he'd taken a photo of *Clumber* on that special April day back in 1961, so I assume that he did not! At the time, I forgot to ask and jumped on a departing train heading for the Epping Forest.

I have very recently been made aware that my original spotting mate John Boulden did end up working as a driver

Class B1 4-6-0 No. 61300 passes through Long Melford station tender first as a Craven Class 105 2-car DMU arrives off the Bury branch in 1961. Photo: Dr. Ian C. Allen (ICA E3825) © The Transport Treasury

on the Eastern Region, Colchester Main Line. He then became a manager for Anglia Railways. He is still around somewhere with an occasional presence on the "East Anglian Rails" Facebook group. We lost touch when I went to university. If anyone reading this article knows him could you please mention that the B17SLT would like to talk to him.

There are witnesses out there other than me of the extended lives of some of the B17s. It would be good if B17SLT members could track them down. I plan to stay around until new-build No. 61673 *Spirit of Sandringham* is no longer a figment of our imagination but a meaningful, functioning hunk of live, steaming metal...

This article is dedicated to Peter Wright (Director and Historian – B17SLT), his favourite B17 is No. 61620 Clumber.

John E. Butcher, B17 Steam Locomotive Trust Membership Number: 0263

I first spotted Paxman Class 15 Type 1 Diesel-Electric D8243 on the Stour Valley Line. It is seen here leaving the suburban lines at Kings Cross. Photo: © The Transport Treasury

61620 *Clumber* pictured working the 12.43pm Ely-Liverpool Street service at Barnwell Junction on 29th July 1953.
Photo: Arthur Carpenter © The Transport Treasury

NO. 60100 SPEARMINT

BY KEVIN POTTS

LNER A3 Pacific No. 2796 *Spearmint* at Dundee Tay Bridge with an Aberdeen to Edinburgh Waverley express in June 1932.
Photo: © T. G. Hepburn/Rail Archive Stephenson

Daring to name just one of Gresley's celebrated A3s as a "favourite locomotive" is bound to involve the inevitable question: "Why?" And why is a Geordie exile currently resident in the South West, and familiar with the locomotives designed by the likes of Bulleid and Maunsell, professing such admiration for a single A3?

The answer lies back in 1968. Being led from a family visit to a relative in Chester-Le-Street, I looked down towards the viaduct carrying the East Coast main line over the town to see a sudden flash of green, a ribbon of white steam and heard the piercing sound of a whistle. Without realising it, I had witnessed No. 4472 *Flying Scotsman* on its epic recreation of the non-stop London to Edinburgh run and a lifelong interest was sparked. Fast forward a few years to find former NBR, LNER and BR Driver Norman McKillop's autobiography 'Enginemen Elite' on a bookshelf and, if *Spearmint* was good enough for Toram Beg, then along with an admiration of her A3 brethren, she was certainly good enough for me.

Norman McKillop joined the North British Railway as an engine cleaner in 1910. He rose through the footplate grades to Driver, based at Edinburgh's prestigious Haymarket Shed, before taking up a literary role producing articles for the British Railways Magazine. He wrote initially under the Gaelic pseudonym Toram Beg (wee Norman) before producing 'How I became an Engine Driver' in 1953, followed by the more detailed 'Enginemen Elite' in 1958.

The latter book recounts how he remembered seeing *Spearmint* when she was new and allocated to Haymarket in the 1930s as No. 2796. This is how she would have looked, resplendent in apple green livery (left) and entrusted with the LNER's crack express services. Both *Spearmint* and *Call Boy* were recorded as being used on the non-stop between Edinburgh and London from 1928 to 1936. During this period the locomotive was paired with a corridor tender.

Spearmint is recorded (left) at Dundee Tay Bridge station in June 1932, working south from Aberdeen to Edinburgh. From an operating perspective, the crew will be faced with a challenge, restarting the heavy train on a set of reverse curves.

At the end of the war, McKillop was driving in Haymarket's No 2 Link. He was also a leading light in the LDC and MIC at the depot. As such, he promoted the argument that locomotive serviceability and performance would improve if senior crews were to be "given" their own locomotive, as had been previous practice. He got his way and was subsequently given a choice of steeds. He could have nominated an A4 but chose instead 60100. His book goes into detail about the sorry state she was in at the time and how he took pride in restoring her back to her former glory. Now running with a standard non-corridor tender (below), the crew here have ensured they've taken on as much coal as possible at Haymarket – there are some serious lumps for the fireman to break up en route if given closer scrutiny!

Now with the BR logo on a high sided tender (above), *Spearmint* speeds through Alnmouth on 11th June 1957. The crew and locomotive look in top form and clearly on top of the job. Justification for McKillop's push for dedicated crews? Interestingly, the Fireman sits relaxed on his seat, watching the road ahead (and presumably the photographer!). But closer inspection reveals he seems quite 'senior' for a Fireman. I would bet that he's the Driver allowing his Fireman to get valuable experience on the regulator. At this stage in 1957 *Spearmint* has yet to be fitted with AWS.

60100 *Spearmint* sits on Haymarket shed (below) with No. 60057 *Ormonde*. Both locomotives have now been fitted with the Kylchap double exhaust and chimney which transformed their abilities in the final years of service. She received this modification in September 1958 and is now also fitted with AWS – see the protective guard behind the front coupling. *Spearmint* was one of the A3s named after famous racehorses, in this case the 1906 Derby winner.

Spearmint spent most of her life working from Haymarket in Scotland. However, here she is (opposite page, top left) working a freight through West Hartlepool on 4th August 1960. Possibly a running in turn after maintenance/overhaul or a balancing working, or she could have been commandeered by an enterprising Running Foreman before Haymarket called her back!

Running under a cautionary aspect from the Distant signal, *Spearmint* (opposite page, centre) comes off the Forth Bridge with a mixed goods. I wonder what McKillop would have made of her appearance now? The softer exhaust produced by the Kylchap exhaust gave rise to problems with drifting smoke and subsequent difficulties in signal sighting. The solution was to fit the smoke deflectors that became (and are still) a topic of conversation that divide lineside opinion as to whether they improve the locomotive's appearance or not. Since this article is about my favourite A3, I'll state that I preferred *Spearmint* without but with the double chimney!

Spearmint (top right) where she belongs – at the head of a passenger train, a gentle haze at the chimney indicating the Fireman has his fire well under control, and a satisfied Driver waiting contentedly for the 'Right Away'.

Spearmint was withdrawn from St Margaret's shed on 19th June 1965, one of the last three surviving A3s. She had run an impressive 1,927,116 miles in her 35 years of service and was cut up in Darlington Works.

I do know the whereabouts of at least one of her nameplates...

A final view (below) of 60100 *Spearmint*. Perhaps not the 'standard view', but full of interest. It's clearly an intermediate station stop; note the depleted coal from the very front of the tender. As yet, no need for the Fireman to 'double shovel' to gain access to coal that hasn't rolled forward as the journey progresses. And with his clean appearance and shoes he takes pride in the job. Meanwhile he waits patiently for the tender tank to top up. Plenty of time during the station stop to contemplate either the task ahead, or more likely he's watching passengers boarding whilst anticipating a pint at the Railway Club when the turn ends!

BROXBOURNE JUNCTION SIGNAL BOX
– SOME THOUGHTS FROM OVER 40 YEARS AGO

BY ROGER GEACH

Broxbourne Junction is situated on the London Liverpool Street to Cambridge line at Hoddesdon Essex Road. Railway-wise it was between the stations of Broxbourne and Royden on the main line and Broxbourne and Rye House on the Hertford East Branch. At the time that I worked there, from January 1979 to December 1981, it was under the Management of the Area Manager Broxbourne in the Liverpool Street Division of the Eastern Region. It controlled the Junction for the Hertford East Line and also the entrance and exit to Rye House Power Station.

While it was a coal fired power station only oil trains went into the complex during my time. These were very spasmodic with more in the winter months then summer.

In addition there was a very busy level crossing here named Essex Road. It did have a low underpass where single file vehicles could pass under the railway but only cars could do this. The level crossing was worked by a gate wheel before barriers replaced this during summer 1981. The box itself was a concrete design and not very pretty with plenty of glass to give a good view of passing trains. It was track circuit block with a 35 lever frame that worked search light style signals. The junction was manual but motor points were installed around 1980 when it was relaid. On the main line the adjacent signal boxes were Broxbourne Station and Roydon. On the branch to Hertford East it was St. Margarets. Communication was by block bells, there were no train describers. During early 1982 Broxbourne Junction was no more, it closed and the work was transferred to Broxbourne Station. Preparation for this work continued through 1981 with the Stratford ST Installation gang carrying out resignalling work. Likewise the old wooden gates were replaced by lifting barriers and the junction relaid with motor points well before the closure. The signal box was not very popular as it was busy and one had to wind the gates many times during a shift. There were far less busy boxes in the area of the same grade (B) so there were very often vacancies there. That was fine if you liked overtime as I did at the time. There were a lot of vacancies for signalmen during the period I worked here within the Liverpool Street Division. They were always short of staff as shift work was disliked by many people and the railway did not pay very well when compared to private industry locally.

Relief signalmen were often my relief as at one point I was the only regular signalman there. It was noticeable that while there were youngsters like myself and older signalman, there were not many at all in their thirties, almost like a generation gap.

In the off-peak period there would be two trains an hour from London to Hertford East and two trains to Bishops Stortford from London along with a Cambridge or King's Lynn express that departed roughly at 35 minutes past the hour for those destinations. It was the same pattern in the reverse direction. Later on a stopping diesel-hauled Cambridge service had been introduced that left London at 5 minutes past the hour on even hours. From Cambridge on the Up road the departure was 39 minutes past the hour even hours. During the peak hours 07.00 to 09.00 and 16.00 to 19.00 hours there were many more service,s some fast and some semi-fast. Electric Multiple Units provided the local service, mainly in the hands of class 305 units based at Ilford. March and Stratford based Class 31 and 37 locos were the most common on the Cambridge and King's Lynn services. Class 47 were more unusual but once the electric train heated stock had been introduced on the Norwich to London route the non ETH 47/0s started to be diagrammed on the King's Lynn services. The odd 47 did of course appear prior to that but they were not as common on passenger work at that time. As well as passenger workings there was a considerable amount of freight traffic that passed from Whitemoor Yard, March to Temple Mills in East London. While March and Stratford class 31 and 37s were the normal power, more exotic types could appear on these workings, where they were out and

May 1979 • Broxbourne junction signal box beside Essex Road, Hoddesdon on an early May morning. Probably a Sunday morning with little traffic about on either the road or railway. Note the ladder leading onto the roof, very close to the overheads, I seem to recall this was removed at some point. One would certainly not want too get close to the overhead wires.

back March drivers, along with foreign depot 31 and 37s. Basically any type of loco March men knew how to drive could appear. Classes 20, 25, 40, 45 even a 46 were all used. The low numbered class 56s were also seen travelling from Shirebrook to Stratford Diesel Repair shops for rectification work during autumn 1979. I believe it was for bogie changes, as they returned quite quick, always as light locos. I wondered who drove them as Stratford Depot did not learn how to operate class 56 locos at that time. The rarest locos I saw were class 50s where a few made their way from the Western Region to Doncaster via Stratford, the Lea Valley and March. Some hauled trains while others ran light loco. Southern Region class 33s also appeared on troop trains to Thetford and return. They also worked excursions to Spalding for the annual flower festival held during the first week of May.

The Whitemoor to Temple Mills freights ran as class 8 at the time, partially fitted to run at 35 mph with a brake van in rear. The only class 9 trains I recall were the Leyton to Chesterton Junction prefab services which were unfitted, thus non-braked. Various fully-fitted services ran with a large number of oil services from North Thameside to a number of destinations which varied quite a bit from week to week.

Many ran via the weekly train plan as short term planning (STP). You received the weekly notices as to when and where they were going by the guard of a passing Hertford East train throwing them out his window into the signal box. One would get a warning phone call that the notices were on a certain service. A few of the regular oil services were as follows. 6L34 19.50 Ripple Lane to Cambridge Charrington oil tanks which was booked to run MWTHO and were often early runners. Two other regular oil services were the 6J44 20.55 Ripple Lane to Kilnhurst service and the 6N46 23.25 Ripple Lane to Norwich tanks.

There were many special freight services in those days conveying all sorts. The concrete works at Lenwade would send out bridge parts which ran as out of gauge loads, often late in the evening. I signalled a number of these that went all over the country and ran via Thetford, Ely, and Cambridge to Temple Mills before going onward to the eventual destination. They ran as special bell signal which

July 1979 • The view of the 35 lever frame and block bells. Only the Roydon one is modern far right. Note the crossing wheel, used to open and close the crossing gates so many times on a twelve hour shift it kept us all very slim and fit. From memory there were nine white levers at the time, indicating not used.

August 1979 • The wooden gates seen late on Sunday evening when the sun has come around. We are looking from Hoddesdon towards Dobbs Weir and the power station ground is behind the trees on the left. Note the red oil LNER lamps on top of each gate.

September 1981 • On the down road looking from Broxbourne towards Broxbourne Junction was the first controlled signal BJ4. Not long left now as the replacement signal complete with route indicator for the Hertford East branch is in position. This will become a BN signal worked from the panel in Broxbourne signal box along with the rest of Broxbourne Junctions signals. BJ5 was the controlling signal for the Junction on the down at the time. With the resignalling the opportunity has been taken to change the location of the signal protecting the Hertford branch Junction, giving more warning.

was 2-6-2, they could not be diverted and were subject to a number of restrictions.

There was a regular freightliner service, the 4C60 02.15 Leeds to Tilbury, due by Broxbourne at 09.40 hours which brought a 47 from just about anywhere, at times even a Western Region based loco. Often there were special freightliners and diversions if the main Chelmsford route was disrupted which happened quite often. A number of special car trains would run from places like Queenborough, Eastleigh and Dagenham.

Chemicals trains, like the Bow Junction to Haverton Hill, was another train that came our way along with cement from Cambridge. Then during the summer special oil trains to Stanstead often from Grain rather than North Thameside. My favourite train was the 8J93 16.10 Whitemoor to Temple Mills, booked Broxbourne at 19.10, but often a foreign loco as the crew returned with 8J94 20.45 Temple Mills to Whitemoor. It was generally well-loaded with a good rake of mixed trucks. Some trains were disliked, like the Croft Original as it was known, the 8E06 13.30 Croft to Leyton conveying ballast for the engineers and often in a motley collection of wagons and the all too frequent hot box. You would have to send 7 bells to stop and examine in those days and the faulty wagon would be shunted out. This would spoil a quiet moment in the late evening and cause disruption. Stop and Examine was more common for passenger services with a door on the catch or in some cases doors wide open! Door interlocking was for the future at this time.

Then there was the local freight, 8P20 09.25 from Temple Mills to Hertford East and return, often hauled by a class 31/0 'Toffee Apple' back in 1979. Due to arrive at Broxbourne at 10.15, it conveyed household coal for Broxbourne Charrington's coal depot. It also had empty wagons for concrete sleepers to be loaded at Costains at Rye House and also Ferry Wagons with wood and tank wagons with glue for Austin's sidings at St. Margarets. The continental glue wagons were for the then Schreiber factory and came via the train ferry. Another local freight was the 7S12 04.52 Temple Mills to Bishops Stortford. The return service, 08.50 departure, was booked to call at Harlow Mill from 09.01 to 09.48 and conveyed sand wagons for the sidings there. The other local freight was empty wagons to load at the scrap yard of Jones and Co. at Brimsdown, the 8P16 09.20 Temple Mills to Broxbourne. We did not see this service as it ran to Broxbourne only then ran round to access Brimsdown from the Up line. In addition to freight traffic there were a number of parcel services running at this time as the National Carriers traffic was still conveyed by rail along with Royal Mail traffic and newspapers as well.

The only quiet shift tended to be a Saturday night and then only if no engineering work was taking place.

The other signalmen were generally friendly and you would be invited to other boxes to see how they worked and enjoy a brew or two. The topic of conversation would

8th December 1981 • We had some very cold winters when I was working here. With snow on the ground Stratford's 47130 passes Broxbourne Junction on the Up with the 13.30 King's Lynn to Liverpool Street service. Note the lifting barriers have replaced the crossing wheel. On the down in the distance is BJ5 signal.

4th July 1980 • 31221 of Tinsley depot makes a change from the Stratford and March based class 31. This was the 12.39 Cambridge to Liverpool Street service which ran fast from Harlow Town to Tottenham Hale. Note 31221 has its number in the headcode box as well as on the side. Class 31s were very common on this line.

6th December 1979 • One of the numerous special automotive services we would see passing by is represented by 37125 on a special from Eastleigh to Wrenthorpe (Wakefield). 37125 was a rare loco for those that sought to be hauled by all the class 37s on passenger services. Very much an Eastern Region freight loco for much of its time it was converted and renumbered 37904 in 1987.

18th February 1981 • March men going home with a special Temple Mills to Whitemoor freight. 20173 and 20198, both Toton based locos, have a single yellow at BJ5, no doubt following a stopping Bishops Stortford EMU. I had a relief man training in the box at the time so popped out for a picture. There were many special freights that ran at that time in addition to or replacing a booked service.

always include how much overtime and Sundays one could grab, and who was making the most on overtime. A Sunday 12-hour shift paid an equivalent of 21 hours pay then. Relief men got paid travelling time from their home station base. No-one wanted to work at ones home station but always elsewhere to book travelling time. We would also discuss Rules and Regulations particularly if there were any amendments to the Rule Book or Signalling Regulations. Any strange locos that had been seen would also be discussed with those who were interested. One of the most remarkable workings was a class 40 that went to Hertford East on the local goods and the relief signalman took pictures to prove it. Some of us would go out together on a rest day for a day out somewhere, maybe visit a preserved railway and have lunch in a local pub, all very sociable. During times of engineering work the drivers and guards would come to the box for a chat and make a cuppa. They were nearly always Stratford drivers and guards, and very often you would get the offer of a footplate ride somewhere on your rest day if you were interested. It was a good way to learn more about the job and most staff were friendly and good company. Likewise some drivers would come and visit the signal box when they were off-duty.

The official turns of duty were 06.00–14.00, 14.00–22.00 and 22.00–06.00. Sundays were 12 hour days, either 06.00–18.00 or 18.00–06.00. The box was manned 24 hours a day throughout the year and only closed on Christmas Day. Often there was a local arrangement and we were relieved at 06.30, 13.30 and 21.30. I have mentioned how the shortage of staff meant 12 hour working regularly as well as some relief men would perhaps be engaged in engineering work on nights, go home at 06.00 and be back to work a box at 14.00 the same day! I would double back after working a 12 hour Sunday night shift, going home at 06.30 and be back at 13.30 hours for the Monday afternoon shift. That was common at the time all over and when young we thought nothing of it. You just had to make certain you had a good supply of tea and grub!

13th May 1981 • Plenty of new built stock came along this railway. Here we have 31234 of Tinsley depot, very clean and hauling new build 315059/060 from York Works to Ilford. Behind we have a good view of Rye House power station. On the down BN5 displays a double yellow for a service towards Roydon. For Hertford East trains this was an approach control signal with a more restrictive speed around the junction. Class 508 units when built at York also came this way hauled to Strawberry Hill via March Cambridge and Temple Mills. Many transit moves came during the night so they were not able to be photographed.

5th October 1979 • 56034 passing the signal box returning from the Diesel Repair Shops at Stratford running light to Shirebrook. Rectification work was carried out on quite a few of the low number Romanian built class 56s at Stratford at this time. I saw quite a few of these locos long before they ever had any regular work in this area. Note the W. J. Simms Hoddesdon-based truck, very much a regular sight here as they would trunk between warehouses over the crossing.

6th April 1979 • A visitor from the Southern Region, 33001 works a Thetford to Ludgershall Troop train special past Broxbourne Junction. I observed a few troop train specials but they were spasmodic and unpredictable as to when they would run.

13th March 1978 • A view of the Broxbourne Station signal box at the north end of the station with a good view over the track and the coal yard. One could often tell who was on duty by the identity of the car parked opposite the signal box by the coal unloading point. Some signalmen did travel by train of course. Stratford's 37039 is passing with an up King's Lynn to Liverpool Street service. The Gasometer stands out very much a local landmark at the time.

21st August 1980 • Roydon, a delightful country station by the River Stort, seen just after 15.00 with Old Oak Commons' 31230 working the 14.36 Liverpool Street to King's Lynn service. The signal box is out of view to the right. Roydon had staggered platforms. It was not unusual for Control to pinch locomotives from other regions when the opportunity arose. However they would also do the same to you but that's another story.

25th May 1981 • No. 55009 Alycidon is seen from the steps of Spellbrook signal box which was situated between Sawbridgeworth and Bishops Stortford on the Liverpool Street to Cambridge line. The occasion was the Deltic Fenman 2 from Finsbury Park to Wansford out via the Lea Valley to Cambridge, Ely and March to Peterborough. Very few Deltics ever worked this route, they were more common between Cambridge and Ely via March when East Coast main line services were diverted due to engineering work.

25th July 1983 • The different front ends on the Class 31 locomotives can be seen here at March depot. Parked up on the stops by Hundred Road are Nos. 31242 and 31215. 31242 was at the time allocated to Immingham and 31215 to Gateshead, moving to Thornaby. Those that recall March will remember the locos were stabled right up by the public road. March provided motive power for the numerous freight workings as well as the Table 18 Norwich to Birmingham passenger services.

EASTERN TIMES • ISSUE 2

B1 AT BANAVIE
Tokens are exchanged as Thompson Class B1 4-6-0 No. 61352 enters Banavie station on the West Highland Line with a service for Fort William.
Photo: Gerald Daniels @ The Transport Treasury

EASTERN TIMES • ISSUE 2

EASTERN TIMES • ISSUE 2

THE LAST SHUNTING HORSE
BY KEVIN ROBERTSON

All photos by Paul Hocquard © The Transport Treasury

The horse as a prime mover had a life of several centuries, likely millennia. Employed as a beast of burden, a means of transport and also for its sheer muscle, has served man well (and still does in some parts of the world) but perhaps not always given the respect and care it deserves.

Despite there being few if any horses in regular commercial use, certainly so far as UK transport is concerned, there are still occasions when a return to olden days may be witnessed with a horse drawn barge, a dray, carriage, and of course shackled to the plough. In some areas though it is certainly defunct, one of these being as a shunting horse on the railway – not that a lot of shunting takes place nowadays anyway – somehow the idea of a horse attached to a modern EMU is not what any of us likely regularly have in mind. Even so the horse does still maintain a connection with the modern railway and that is when the term 'horse-power' is mentioned. 'X' number of horse power meaning literally that, a multiple of the power that number of horses might achieve – man at best only able to muster a maximum half of one horse power. Hence the connection between the living horse and the modern machine but no so the other way around; for whilst we also relate the terms 'tractive effort' and 'torque' when it comes to locomotives and units these are not really transposed to the horse. To digress, in steam days we might also have referred to 'coal per mile', the equivalent translation being 'miles per bale of hay'.

Away from both generalisation and the whimsical, horses were associated with railways from the earliest days. For construction purposes and then for hauling some passenger trains, for shunting and of course for deliveries. Railway-owned horses might well be stabled – again the term remains in use for locomotives and rolling stock – although no doubt on occasions the local coal merchants horse might also be borrowed to move a wagon or similar if a locomotive were not available.

With thousands of horses on their books, the railway companies similarly had well organised facilities for stabling, fodder, as well as the services of farriers and vets; in 1913 a countrywide figure of 21,826 animals is quoted, the majority of these engaged in pulling carts on local deliveries. This number reduced to 9,077 in 1945 and 9,000 at nationalisation three years later. (It was just 60 eleven years later in 1959.)

As an indication of how the wheel has turned – no deliberate pun intended – with the horse vis-à-vis the motor vehicle, we need first look back to London at the turn of the 19th/20th century and where the associate manure that covered the streets was considered a health hazard, at the time more so than the fumes from the 'modern' motor vehicles already rapidly coming on the scene as replacements. This falling out of love with the horse continued through the early decades of the 20th century, in effect a relentless war of words against horse transport. Indeed the district goods manager of the GWR in Birmingham informed the West Midland Traffic Commissioners in 1936 that Birmingham City Council was strongly behind replacing horses with motor vehicles, quoting that, "before long it would be necessary to compel railway companies to take horses off the central streets."

The disadvantages of the horse when compared with mechanised alternatives were well described in a Manchester Guardian article of 1952. Here it was stated the Road Transport Division of British Railways had set up an experiment where 100 drivers were allocated 'electric horses' rather than the 'leg in each corner' type. Hull was one of the stations that took part, with 12 of the new machines. What was presumably a registration number, YE 4102 was quoted and a particular driver, Charlie Pulford who forsook his horse 'Tiny' for the new motor.

When interviewed Tiny naturally looked at it from his own point of view. Quoting the newspaper reporter, 'Charlie Pulford thinks that the only advantage Tiny had was that he knew his own name and would come when you whistled, whereas YE 4102 does not.'

The article continued by singing the praises of the motor. It did not need to be stabled, groomed, mucked out, fed, or require the services of the farrier and harness makers. The only maintenance Charlie appeared to think that YE 4102 needed was a quick wash with the hosepipe, and which showed an innocent and touching faith in the machine's reliability. Mr. A. A. Harrison of British Railways argued that the 'electric horses' saved the country petrol and also cut the use of manpower between 30-60% (the figures and grades involved are not explained). The electric horse could also be recharged during the evening, therefore not interfering with the demands of industry. (Where have we heard similar arguments about the benefits of electric vehicles more recently...?) The Manchester Guardian did not shy away from one final advantage of the electric horse: it didn't involve you in a moral conundrum when the time came to pension it off; the latter perhaps part of a comment from another unnamed railwayman who observed, "...at least with the electric horses when it comes time for them to be pensioned off, there will not be one group trying to put them out to pasture, and another trying to eat them."

Indeed aside from natural causes the prospect of a working horse going to slaughter was a real one. In WW1 thousands of horses conscripted into military service in France were slaughtered after 1918 rather than bringing them home. Post WW2 the number of horses in Britain fell probably even more drastically as mechanism became commonplace. As an example in the annual report from what is now the Blue Cross charity in 1952 said that 719,500 farm horses had been slaughtered since 1939, and estimated that the equine population had reduced by 1.5 million over the previous 14 years. The Blue Cross did however, single out British Railways for praise, as it had agreed to sell as many of its redundant horses to various charities as could afford to be bought. The public it appeared, had a special affection for the railway horse. For many people in towns, the agricultural horse was a distant creature, not often seen, but the railway horse was different. It delivered goods to their workplaces. It delivered parcels to their door, consequently they fed Tom, or Ben, or Kitty, as they stopped on their routes. They were part of everyday life.

Away from the LNER we might single out the example of Bobby the last surviving shunting horse at Blackburn, made redundant – perhaps we should use the railway term 'withdrawn' – in 1954, as BR were no longer prepared to cover his £1 5s 0d daily 'maintenance costs'. The official statement saying he was 'no longer cost effective'. Bobby's fate was not recorded.

Perhaps it was the story of Bobby, reported in the local newspaper, that stirred the soul of the nation as a whole, a nation we are told of animal lovers. Local newspapers printed story after story describing vigorous local campaigns to save the railway horses of their towns and cities. There was also the Dumb Friends' League which set up a lease and lend scheme. They, with the public's help, would buy railway horses, and then re-home them, with regular inspections to ensure the horses' welfare.

On 29th January 1954, the Northampton Mercury reported the story of an anonymous local businessman (described as 'the owner of a very small business at the end of a back street') who had heard that the nine horses at the town's Castle station were to be replaced with lorries, and might end up in the slaughterhouse. He contacted the Blue Cross and Our Dumb Friends' League and together they started a campaign to raise the £540 to buy the horses; by 16th February, the target had been reached.

This was followed in August, when the same newspaper printed a description of the last days of Northampton's railway horses, and their journey to retirement on a farm in King's Sutton. It was also ironic that horses who had spent their entire lives transporting things were terrified of being transported themselves. After years of work at smoky Castle station, this was a new experience for Joe and Ben. First they munched the thick grass, and then, realising they had more space than they had ever seen before in their lives, kicked up their heels and galloped off together in sunshine.

Happy though they were, they had not been too keen to leave Castle station earlier in the day. It took fifteen minutes to load Ben into the Blue Cross horse ambulance in which they travelled, and twice as long to coax an extremely reluctant Joe. So reluctant was Joe to leave that once he broke away and ran back to his stables, shivering with nervousness.

We should also not forget the effect on the former railway staff. It must have been extraordinarily difficult for Harry Hawtin, the stable foreman at Northampton who was losing his old and trusted friends. What happened to Harry is not related, we can only hope he was deployed elsewhere on the railways, there being few public collections to help redundant railwaymen.'

Being 'horsey' also had no bearing on people's willingness to save the horses. Mrs. Ann Newton recalls her interest in railway horses was sparked when she gave an ice cream to a piebald railway horse in Leeds who then called every day for a tit-bit. When he was sent to auction in Manchester, Mrs. Newton bought him, and from that point on devoted herself to working along with another (unnamed) charity to save the railway horses of Leeds. Mrs. Newton also cut a distinctive figure. No tweedy attire, in 1952 the Yorkshire Post and Leeds Intelligencer described her thus, 'The tough horse dealers and their hangers-on in a Manchester auction yard are now used to seeing the spruce figure of Mrs. Newton, always looking as if she had come straight from a London fashion show, elbowing her way through the crowd to bid for a horse against stiff competition.

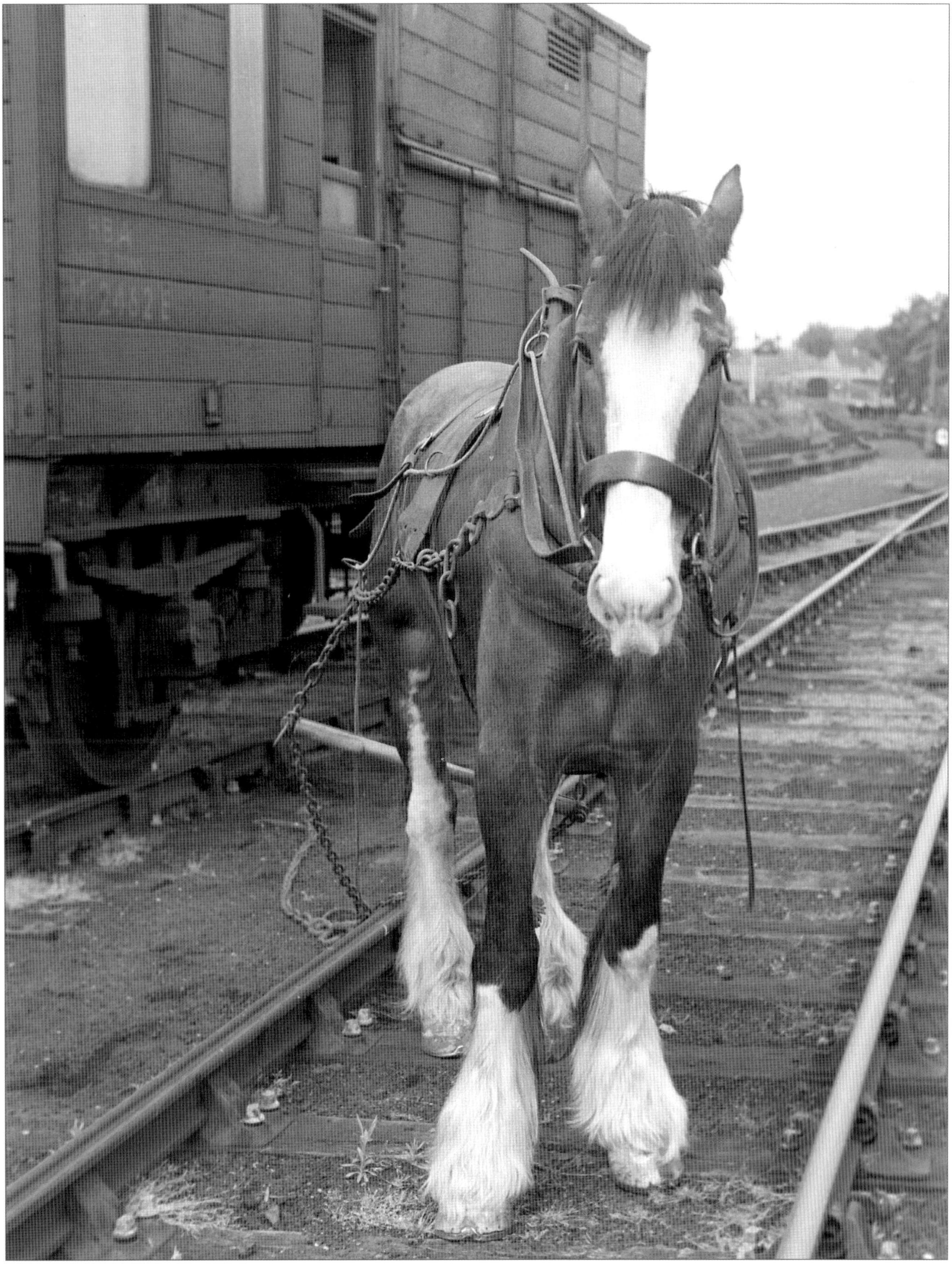

She never 'dresses down' to go to the auctions – sometimes she wears an even more daring hat than usual. She is an incongruous figure in the gloomy shed, filled as it is with frightened horses and with shouting and the cracking of whips.' Ann Newton raised enough money to save several of the Manchester horses, and many others throughout the north.

Even so the public's enthusiasm for saving horses did not always meet with unmixed joy from railway staff. For example when a fund was started in 1952 to save the redundant horses in Rochdale, and the Manchester Guardian reported one railwayman as being 'fed up with shoving hundreds of people, including children, round the stables.'

Other railwayman though were proud of their steeds. Many of the detailed branch line history books describing and sometimes illustrating a shunting or dray horse from long ago often complete with brass decorations. Indeed there was a time when the railwaymen and their families would willingly groom the horse to perfection and regularly enter the animal in various shows.

Around the country then former railway stables were rapidly becoming redundant, some demolished and replacement buildings erected around their foot (hoof) print, whilst others were simply left to rot or perhaps turned over to new uses.

There would have to be a final one of course and that appropriately enough was at Newmarket, a town famed for its racing connections. It was here in 1960 that were stabled the last three British Railways shunting horses, Tetley, Butch and Charlie. Ironic too as all three spent some of their time shunting horseboxes containing their upmarket racehorse cousins. Tetley was first to go, followed by Butch and so leaving just Charlie who naturally became somewhat of a celebrity, visited by Pathé News as well as various journalists and renowned photographers. One of the latter was Paul Hocquard whose work accompanies this piece, although regretfully without contemporary caption notes.

At the time of his retirement on 21st February 1967 Charlie was 20 years of age being cared for by railwayman Tommy Butch who is also seen in some of the images. Ironically Charlie's very last task was to shunt a horsebox on to a train in which he would then travel into retirement. This was to Clare Hall, Stone Easton where he would also be reunited with his erstwhile stable companion, Butch.

The railway horses of Britain felt their way into the public consciousness in a way their mechanised replacements could not. However inconvenient and overly labour-intensive the horse came to be seen, the opportunity they gave the public to connect with another living being, to interact with an animal that was pleased to see you even if it was just because you gave it an ice cream, led to furious fights to save them.

REFERENCES

Our, S. C. (1952, October 22nd). RAILWAYS TRADE "TINY" FOR AN ELECTRIC HORSE. The Manchester Guardian (1901-1959) Retrieved from http://search.proquest.com/docview/479404792?accountid=55962

1,500,000 FEWER HORSES. (1952, 3rd Nov). The Manchester Guardian (1901-1959) Retrieved from http://search.proquest.com/docview/479394091?accountid=55962

Riding Magazine, August 1951, pg 313

Northampton Mercury – Friday 29th January 1954, Image © Johnston Press plc. Image created courtesy of THE BRITISH LIBRARY BOARD.

Copyright Northampton Mercury – Friday 13th August 1954, Image © Johnston Press plc. Image © Johnston Press plc. Image created courtesy of THE BRITISH LIBRARY BOARD.

Yorkshire Evening Post – Thursday 14th August 1952, Image © Johnston Press plc. Image created courtesy of THE BRITISH LIBRARY BOARD.

FUND GATHERS £180 TO SAVE HORSES. (1952, June 28th). The Manchester Guardian (1901-1959) Retrieved from http://search.proquest.com/docview/479357825?accountid=55962

The Railway magazine – various issues.

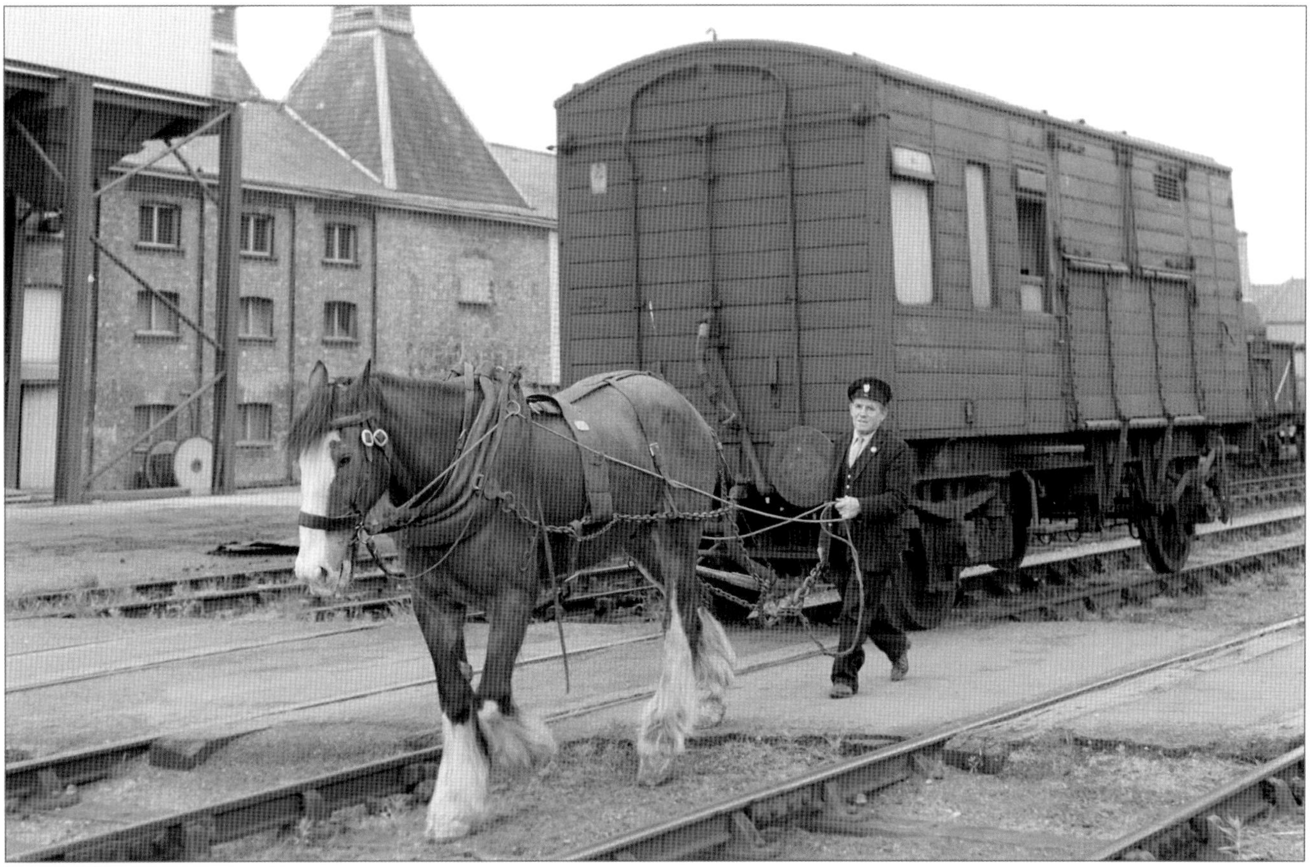

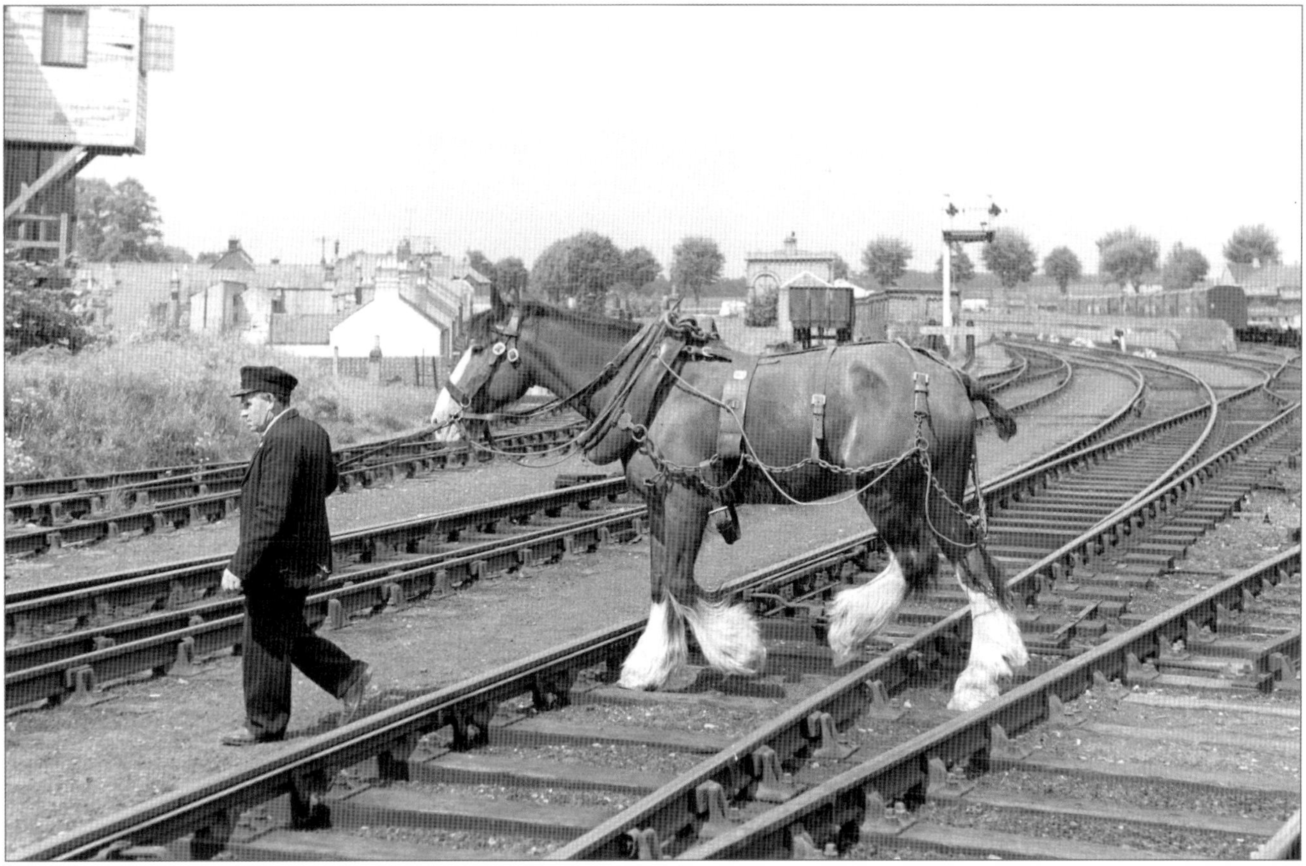

THE 1935 LNER NEW WORKS SCHEME – STRATFORD TO ILFORD
A PHOTOGRAPHIC ARCHIVE

BY DAVE BRENNAND

Overcrowded, steam operated passenger services in East London and Essex became a major headache for both the London Passenger Transport Board (formed in 1933) and the London North Eastern Railway in the late 1920s and early 1930s. London's suburbs were expanding with new housing development putting a severe strain on the capital's transport system.

Antiquated motive power, rolling stock, signalling and infrastructure desperately needed modernising through massive investment. This came partly from the Government in the shape of subsidised loans to the railway companies in order that major engineering works were created in areas of high unemployment. Thus was born the 1935-40 London Railways New Works Programme which saw colossal investment in the LPTB Bakerloo, Northern and Central lines, with new trains, stations and tunnels east of Liverpool Street to Stratford and further tunnels to connect with the LNER steam operated Epping, Ongar and Hainault Loop services. The Central Line extension from Liverpool Street to Stratford in new tunnels started in the late 1930s, coming up for air at Stratford to provide a new interchange with the surface lines, before plummeting into tunnel again and resurfacing at Leyton, bypassing the former slower surface route via Loughton Branch Junction near Temple Mills Lane. At Leytonstone a new route was tunnelled to Wanstead, Redbridge and Gants Hill, thereby connecting with the former Seven Kings to Newbury Park line. During the war, the tunnels at Bethnal Green were used as air raid shelters and the tunnels in the Redbridge to Gants Hill area were used as a Plessey factory contributing to the war effort.

The LNER line from Liverpool Street to Shenfield would see electrification, resignalling, track alterations and new rolling stock, which must have given hope to the poor long-suffering commuters. All these plans were thwarted by the outbreak of the Second World War in 1939, but not before some of the work was well underway. Work recommenced upon the cessation of hostilities in 1945 and the construction work took another two to three years to complete.

By very good fortune, the LNER Civil Engineers Department sent out photographers to record the building works in East London, and the pictures accompanying this article give an insight into the upheaval endured over many years by the travelling public. To the east, a new flyover was built at Ilford to swap trains on the Main and Electric lines. To alleviate some of the congestion at Liverpool Street, suburban trains from Essex and East London were dealt with on the East Side platforms 16-18, whilst main line services occupied the central platforms 9-15. As will be seen in the photos, Maryland Point, Forest Gate and Manor Park only had two platforms for local services prior to the New Works Programme.

Upon completion, these stations were expanded with platforms serving all four lines. The former GER stations, dating back to 1865 on the acquired Central Line's Epping, Ongar and Hainault Loop sections and the New Works improvements are just as interesting as the LNER Stratford to Ilford section, but this article is formed around the accompanying photos, some of which are very rare and give an insight into a bygone world. The largest structure on this route was, and still is, the Ilford Flyover. A landmark that I have admired for most of my life. Being raised in Manor Park, I would spend many hours as a youngster peering through the railings alongside the footpath which separates the railway from the City of London Cemetery.

The focus of my attention would always be the endless succession of freight and passenger trains, which shaped my lifelong love of railways and sharing images from the past.

Left: Although undated, this is believed to be a postwar view of Stratford looking towards London from Platform 8. Closer inspection reveals that the overhead gantries for the electrification are in place, but they await the installation of the catenary system. The delightful joint LNER/Central Line poster proclaims the benefits of the completed programme which had been so heavily delayed by the Second World War. A new cream and blue tiled advertising hoarding is in place and beyond is a rustic pre-grouping timekeepers box. A small brick built structure at this exact location provided some shelter to several generations of trainspotters from the biting winter wind. *Photo: Andy Grimmett Collection*

May 1940 • The Central Line tunnels from Liverpool Street had reached Stratford by the time of this view taken in May 1940. The Up Electric line entering the new platform 5 is sandwiched between the new Central line inclines towards Leyton. In the middle is Stratford Central Junction signal box standing on the former platform 1, which was destined to become the new platform 8. This work was given great priority as having just three platforms to cope with all the suburban and main line traffic would have been unimaginably difficult in the rush hours. To the left is the Eastern Curve which joined the North Woolwich branch, although this short link never carried timetabled passenger trains. It is thought that this image was taken from a temporary signal box.

May 1940 • A fine view of the rebuilding at Stratford from the London end looking east. The ill-fated bay platforms 4 and 7 are seen being built. These would have been for the Fenchurch Street shuttle services proposed under the electrification scheme. They were never used for their intended purpose, but platform 4 was eventually acquired by the Docklands Light Railway. The old lamp standards have been replanted on the new platform; this was wartime after all.

1938 • The planned shuttle service to Fenchurch Street saw the need for an additional bi-directional line between Stratford and Bow Junction. A new retaining wall is being constructed in this view of the steep incline to Cook's siding looking towards Bow Junction in 1938. T. Wordsell Class J15 0 6 0 No. 7917 is waiting for the signal prior to attacking the incline. The factories and gas works in this area produced some truly obnoxious pollution at this time.

1946 • The war is over and Europe has been liberated. The year is 1946 and the new Central line tunnel walls, built before the war are already grimy despite not carrying trains at this stage; another indication of the polluted atmosphere. In the foreground is the trackless bay platform 7 and in the distance is Western Junction signal box controlling the curve to the North Woolwich line on the left. A wonderful array of mainly lower quadrant signals would soon be swept away and replaced by colour light signals.

October 1946 • A delightful contrasting view eastwards at Stratford in 1946, showing the grimy past on the left with a steam hauled freight passing through a platform adorned with gas lighting and enamel adverts framed by a wonderful array of semaphore signals. On the right is the new platform 8 with modern canopies under construction, awaiting the installation of overhead electrification and the eagerly anticipated new electric trains.

1936 • Viewed from the London end this is Maryland Point Junction in 1935. The Slow lines are on the left and the Fast lines are on the right. For such a minor station, the GER spent a huge sum of money providing an entrance building at either end of the station. The one seen here was in Water Lane and an even larger entrance hall was also provided in Leytonstone Road. The signal box closed on the 29th June 1946.

1940 • Both of the fine GER entrance and ticket halls at Maryland Point were sacrificed as part of the GE electrification scheme. A new centralised booking hall was built with the once familiar cream and blue tiles commonly seen at Stratford, Forest Gate and Manor Park. This view is looking towards Stratford in June 1940. The suffix 'Point' was dropped from the station name in 1940, but some signs with the original name on were still in situ in the early 1970s. *Photo: J. E. Connor Collection*

1936 • The Great Eastern Railway once again lavishly provided two entrances at Forest Gate. This is the lesser known original building in Forest Lane, captured in 1936 and closed in 1940. The much larger entrance and booking hall in Woodgrange Road opened in 1870. This was rebuilt in 1894 when the line was quadrupled. The right hand side can be see straddling the Fast Lines. It has been sympathetically restored and now serves the Elizabeth Line. The signal box closed in 1946.

1946 • A view looking east at Forest Gate after WW2, with demolition underway for the new four platform station. The old nameboard proclaims 'Forest Gate for Upton' which was either to help arriving passengers or poach customers from the LMS station at Upton Park well over a mile away to the south. These are the Slow Lines which would become the Fast Lines after the rebuilding. Most of the retaining wall survives.
Photo: J. E. Connor Collection

1938 • The incredible pride that railwaymen had is in evidence here. A well kept Manor Park station is seen prior to the New Works rebuilding in 1938. Platelayers attend to the trackwork with the same level of pride. Competitions were held and prizes given out for 'The Prize Length' of track, something that has dissapeared over the decades, as most trackwork is maintained by gigantic tamping machines today. We are looking east. The retaining wall on the right survives.

1936 • Manor Park station in 1936 shows the simple two platform arrangement with the fast lines on the left and slow lines on the right. This would be reversed by the tracks crossing over each other at Ilford on the new flyover. The building dates from 1894 and was damaged by a German bomb on the 24th July 1944. The signal box closed on the 18th August 1949; its duties being shared between the new Forest Gate Junction and Ilford Station boxes.

1935 • This view shows Manor Park goods yard looking towards Ilford in 1935. The yard was still in use for coal traffic until the mid-1970s and even today it is possible to see the space it once occupied. In the distance is Rabbits Road bridge and a small signal box controlled the country end points for the Up Goods Loop and the goods yard.

Circa 1938 • A westward view circa 1938 from Aldersbrook sidings towards Rabbits Road bridge showing early site clearance work for Ilford Flyover. The brick built piers were erected the following year, before being abandoned on the outbreak of war on the 3rd September 1939. They stood for six years untouched and work recommenced in 1945.

Circa 1938 • Another 1938 view at Aldersbrook looking east towards Ilford station showing early construction of the flyover piers. Note the three style of railway seen here. The standard gauge for regular trains, a small section of similar gauge track for the antiquated crane and a narrow gauge railway for tippler wagons removing spoil. The signal box nameboard proclaims ILFORD ALDERSBROOK, which surprised some Signalling Record Society members who believed it was just called Aldersbrook!

May 1939 • Just one year later we see the cast iron superstructure supporting the central section of the flyover in place. Unlike the previous view, Ilford station can be seen in the distance. The same crane is amazingly still in one piece; a testament to the make do and mend construction techniques of the 1930s. That said, the iconic flyover is standing unscathed today and has carried many millions of trains since 1948. It has seen steam, diesel, electric and even the latest bi-mode traction during its existence.

5th May 1935 • Just four months before the outbreak of war on the 3rd September 1939, we see two LNER services passing the half built piers. In the background is the City of London Cemetery, which opened in 1856 and covers some 200 acres. An eastbound express is hauled by B17 4-6-0 No. 2802 *Walsingham*. Being rapidly overtaken is N7 0-6-2T No. 870 hauling ancient suburban stock. The flyover remained in this condition throughout the war.

1949 • In 1949 postwar East London and Essex the brand new AM6/306 3-car electric units were warmly welcomed by commuters who had suffered years of bombing raids, rationing, destruction and terrible loss of life. This is the first unit standing in Ilford Car Sheds in 1949. Designed originally by the LNER, they were very robust, with rapid acceleration, comfortable seats, good brakes and lighting. A monumental leap from the wooden bodied steam hauled stock used in the interwar period.

1930s • We end our journey with a 1930s view of Ilford Carriage Sidings signal box, which also controlled the curve to Newbury Park, made redundant by the opening of the new Wanstead to Newbury Park Central Line. Seven Kings station is in the background and the vast area to the left was transformed in the late 1940s to create Ilford Car Sheds. A second huge building known as the New Shed was opened in the late 1950s to coincide with the arrival of new units for the Southend Line electrification.

THE MAN FROM THE PRU – PART 1
BY PAUL KING

Henry Casserley, and to a lesser extent his son, Richard, are, arguably, two of the more prolific railway photographers of the 20th century. Henry was renowned for his photographs of locomotives whereas Richard appeared more interested in railway infrastructure and rolling stock. The man from the Pru is Henry Casserley, he worked in the Prudential Assurance Company's head office in London all his working life.

Henry died in 1991 and Richard in 2017. After due consideration, the Casserley family took the decision to sell their collection of photographs and negatives. We are all familiar with their work and I had contacted Richard about using some of his father's work and he, considerately, allowed me to borrow some of the photographs taken in my area to scan and a few years later his wife sent some more. The opportunity of acquiring the negatives for the Lincolnshire area was too good to miss and I attended the first auction in Pershore and the second online, managing to acquire all but around half a dozen of the negatives I wanted. Along with the catalogue for the auctions, the auctioneers also provided copies of the negative diaries. These provided a fascinating insight into the Casserley's dedication to recording the railway scene.

As well as making regular visits to locations up and down the British Isles and Ireland, there was also an annual holiday where Henry, later with Richard, spent two weeks travelling around different parts of the country. This article looks at one such holiday taken between 17th and 29th April 1954. The intention was to visit some of the lesser used railway and branch lines in Northern England. The record they have left gives an insight into what has been lost and what has survived, not only locos but stations and rolling stock. The family lived in Berkhamsted, on the West Coast main line, from where Henry and Richard headed north to Rugby where the first photograph of the holiday was taken, ex-Cheshire Lines coach No. 19946.

From Rugby they travelled north to Stoke-on-Trent and then the ex-North Staffordshire line towards Uttoxeter as far as Cresswell for a journey along the branch line to Cheadle before travelling on to Uttoxeter. From there the route took them north towards Macclesfield and Manchester on the 3.00pm departure behind Fowler 2-6-4T No. 42319, photographing the stations at Denstone, Alton Towers, and Cheddleton plus Rudyard Lake from the train. Arriving at Manchester Mayfield, where another Fowler 2-6-4T, No. 42356, was waiting departure with the 5.22pm to Stoke, they crossed the city centre to Victoria before taking a train for Bolton to visit the former, soon to be closed, LNWR shed at Plodder Lane. Two locos were photographed, 4F 0-6-0s Nos. 44384 and 44473 plus an ex-Caledonian Railway coach and an ex-Taff Vale Railway van. This ended the first day.

Returning to Manchester for the night, the next morning found them at London Road from where they caught the 9.28am to Macclesfield Central hauled by Class C14 4-4-2T No. 67447 via Hyde Junction, Hyde Central, Woodley, Romiley, High Lane, Middlewood Higher, Higher Poynton and Bollington. The route took them over ex-Great Central lines as far as Hyde Junction and then the GC & Midland Joint to Romiley and then the GC & NS Joint to Macclesfield. Returning to Manchester on the 11.00am departure behind the same loco. At High Lane they passed Class C13 No. 67433 heading towards Macclesfield with the 10.30am from London Road.

Presumably detraining at Ashburys, they called at the ex-Midland Railway shed at Belle Vue and the GC shed at Gorton. Unfortunately, there isn't a detailed record of the locos on shed at either location. However, Belle Vue yielded photographs of ex-MR 2F 0-6-0 No. 58128 and 1F 0-6-0T No. 41814 together with Black 5 No. 45031 and 4F 0-6-0 No. 44486. J11 0-6-0s Nos. 64294, 64413 and J11/3 No. 64427 plus O4/3 2-8-0 No. 63835 were amongst the locos at Gorton with C14 4-4-2T No. 67447 passing on a train for Macclesfield. On the return to London Road they again called at Belle Vue, photographing 44486 again plus 8F 2-8-0 No. 48224, an ex-M.R. van and the station. At Ashburys they recorded No. 67433 with the 12.48pm London Road-Macclesfield and No. 67447 returning with the 1.20pm from Macclesfield.

At London Road there is reference to a B12 4-6-0, No. 61514, this was a Yarmouth South Town loco at the time and hadn't been in the works at Gorton so what it was doing there is a mystery, unless it has been recorded incorrectly.

18th April 1954 • The visit to Gorton enabled this photograph of Class J11/3 0-6-0 64427 to be taken adjacent to the ash pits.

Fowler 2-6-4Ts Nos. 42322 and 42351 plus Royal Scot No. 46157 *The Royal Artilleryman* were also in the station. A trip to Hadfield was next on the agenda with a return to Manchester on the 4.00pm via Glossop, Mottram, Broadbottom and Guide Bridge hauled by Class C13 No. 67422. It was now early evening but there was still time for a visit to the ex-LYR shed at Newton Heath. Once again there is little detail of the locos on shed, however, ex L&YR 0-6-0s Nos. 52108, 52390 and 52455 together with 0-6-0STs Nos. 51470 and 51472 are recorded plus Compound 4-4-0 No. 41186, Fairburn No. 42283 and Stanier No. 42622, both 2-6-4 tanks, Ivatt 2-6-0 No. 46484 and Fowler 0-8-0 No. 49570. They also photographed several pre-grouping coaches of Furness, Lancashire & Yorkshire, London & North Western and Midland Railway origins and LMS twin coaches 52518 and 52519.

19th April was a Monday and the 9.55am to Manchester Central via Oldham from Manchester Victoria was the next venture, hauled by Stanier 2-6-4T No. 42624. The latter part of the route from Rochdale was over much of the trackbed of the current Manchester Metrolink tramway through New Hey (now Newhey) and Shaw & Crompton. A trip along the short branch from Royton Junction to Royton was made before continuing to Manchester Central behind Fairburn 2-6-4T No. 42279.

The ex-LNWR shed at Longsight and a trip to Stalybridge and Oldham (Mumps) via Lees were next on the agenda. At Longsight, where you would expect to find plenty of larger locos, only Black 5 No. 44687, one of the pair with Caprotti valve gear, driving axles with Skefko roller bearings and double chimney, fitted that category. Other locos photographed were LMS 0-4-4T No. 41906, 2-6-4Ts Nos. 42381 and 42478, Fowler 2-6-0 No. 42886 and ex-MR 0-6-0s Nos. 43275 and 43457. The former L&NWR shed at Lees (Oldham) yielded Fowler 2-6-2T No. 40057 and 0-8-0 No. 49618 plus ex L&Y 0-6-0 No. 52099. After Stalybridge, the train passed through Greenfield, with a trip on the short branch to Delph, Lees, Oldham (Mumps), Parkbridge and Ashton (Oldham Road) to Stockport.

Manchester Central was the starting point on 20th April for a trip to Birkenhead via Widnes Central. Before departure several Cheshire Lines Committee coaches were seen along with Director Class D10 4-4-0 No. 62656 *Sir Clement Royds*, Improved Director No. 62664 *Princess Mary* and J10 0-6-0 No. 65166. At Birkenhead both GWR and LMS types were noted on shed and included GWR Class 14xx No. 1457, 2021 class pannier tanks Nos. 2108 and 2112, 2-6-2Ts Nos. 4122, 4125, 4126 and 4129, 28xx 2-8-0 No. 2822 and 47xx No. 4700. Of LMS origin were 2-6-2Ts Nos 40072, 40101 and 40131, 1F 0-6-0T No. 41734, 2-6-4T No. 42583, Stanier 2-6-0 No. 42969, 0-4-0ST No. 47007 and Swindon built 8F No. 48455. A visit to Hamilton Square and Woodside stations were next, followed by a journey on the former Birkenhead Railway to Hooton, Heswall and West Kirby with ex-GWR types providing the motive power, these included two of the 2-6-2Ts seen earlier at Birkenhead, Nos. 4122 and 4125 plus classmate No. 4120 and 0-4-2T No. 1457. Crossing the Mersey, Hunts Cross, Knotty Ash and Aintree produced J10 No. 65142, Fairburn 2-6-4T No. 42064, ex-LYR 0-6-0 No. 52412 and Hughes/Fowler 2-6-0 No. 42732. The focus of interest at Kirkdale again being coaches where they found more pre-grouping vehicles including ex-Caledonian, Highland and North British examples. Ex-LYR 0-4-0ST No. 51206 was also seen from the train as it passed Bank Hall shed.

The morning of 21st April found them venturing into West Yorkshire on the 9.05am to Normanton hauled by ex-MR 4-4-0 No. 40552. The ex-L&YR shed at Mirfield produced photographs of only one LYR loco, 0-6-0ST No. 51424, an Aspinall rebuild of a Barton-Wright 0-6-0 plus 2-6-4T No. 42411, 2-6-0s Nos. 42856 and 42862 and Austerity 2-8-0 No. 90591. Approaching the station they recorded another

20th April 1954 • Improved Director Class D11/1 4-4-0 62664 *Princess Mary* is heading a stopping train out of Manchester Central on the 20th April 1954. The Manchester Metrolink now runs along the other side of the girder wall on its climb from St. Peter's Square to Deansgate.

Fowler 2-6-4T No. 42412, 2-6-0 No. 42706 and B1 No. 61339. Leaving Mirfield they travelled along the LYR line to Halifax via Brighouse and Greetland then taking the ex-GN, originally Halifax & Ovenden Joint Railway, line from Halifax (North Bridge) to Bradford (Exchange) via Queensbury and Clayton. At Bradford (Exchange) were two of the ageing ex-GNR N1 0-6-2Ts Nos. 69434 and 69450, a third member of the class, No. 69430, was on station pilot duty at Leeds (Central), their next point of call. From Leeds their journey took them north to Wetherby via Collingham Bridge then through Thorpe Arch, Tadcaster and Church Fenton back to Leeds (City), where they caught the 6.35pm to Goole, and an overnight stay, hauled by Stanier 2-6-4T No. 42477, passing Methley and Knottingley (L&Y) on the way.

The following day was to be spent, mostly, in north west Yorkshire but started with the short journey to Stainforth and a visit to Hatfield Colliery. Amongst the locos recorded here were two ex-GWR locos, Hatfield Nos. 3 and 4, and an ex-GSWR loco, No. 5. No.3 was an 0-6-0ST originally built for the Taff Vale Railway and arrived at Hatfield as GWR No. 791, whilst No.4 was an Avonside 0-6-0T built as *Great Mountain* for the Llanelly & Mynydd Mawr Railway and arrived in 1928 as GWR No. 944. This loco survived until 1964 at least. The ex-GSWR loco came to Hatfield after withdrawal by the LMS as 2F 0-6-0T No. 16378. Classmate 16379 also went to a colliery for further service, Hafod Colliery, Denbigh and became the only ex-GSWR locomotive to survive into preservation. The next note shows them at Leeds where they caught a train to Skipton. I have no evidence, but it could well have been the 9.15am Cleethorpes-Leeds which used the Stainforth Junction to Adwick Junction line avoiding reversal in Doncaster station that they caught. Passing through Stainforth at the time were K2 2-6-0 No. 61730 and J11 No. 64410, although there is no record of what trains they were working.

At Skipton where several LMS types including 2-6-2Ts Nos. 40117 and 41325 plus Compound No. 41065 and 0-6-0Ts Nos. 41855 and 47427 were recorded, but the most interesting was a passing freight headed by ex-NER B16 4-6-0 No. 61413 of Neville Hill shed piloted by 4F 0-6-0

20th April 1954 • Taken from the train as it passed Bank Hall shed on its way to Kirkdale, the opportunity was taken to photograph ex-L&Y 0-4-0ST No. 51206 with two classmates, one of which is 51231.

21st April 1954 • Originally built for suburban work out of Kings Cross, fitted with condensing gear for working through the tunnels to Moorgate, the doyen of the ex-GNR Class N1 0-6-2Ts, No. 69430, is captured later in its career at Leeds (Central) station on 21st April 1954. The condensing gear was removed in 1927 when it moved north to Ardsley and from 1941 until withdrawal in 1956 was based at Copley Hill, the GNR shed in Leeds.

22nd April 1954 • The erstwhile G&SWR 0-6-0T No. 323 (LMS 16378) as Hatfield No.5 at the colliery.

No. 44586 from Stourton. The 3.12pm Leeds-Morecambe produced Stanier 4-6-0 No. 44842 which they took as far as Lancaster (Green Ayre). The return journey was from Lancaster (Castle) via Carnforth and the Furness & Midland Joint line to Wennington and thence on to Skipton. This is a presumption as there is no record of photographs being taken between Carnforth and Wennington, although the Richard Casserley records are missing for this period. Noted at Lancaster (Castle) were a Fowler 4-4-0 No. 40362 and Clan 4-6-2 No. 72000 *Clan Buchanan*, whilst at Carnforth where two further Fowler 2-6-4Ts Nos. 42392 and 42393.

Friday 23rd April was to be spent travelling around some of the more obscure lines in East Lancashire. From Skipton they took a train to Earby for the Barnoldswick branch, which was being worked by Ivatt 2-6-2T No. 41273. The next stop was Accrington reached via a circuitous route from Nelson and Burnley Central to Padiham, Simonstone, Great Harwood and Blackburn. A visit to the loco shed recorded ex-L&Y 0-6-0STs Nos. 51390 and 51474, 2P No. 40562, Compound No. 41085 and BR 4-6-0 75045. L&YR 0-4-4T No. 713, withdrawn before the grouping, was also noted as being in use as a carriage warmer. From Accrington the next stop was via Haslingden to Ramsbottom, here they changed for Rawtenstall and the line to Bacup via Clough Fold. Bacup had been the terminus of two lines, that from Ramsbottom via Rawtenstall, over which they travelled plus the line from Rochdale via Broadley and Facit. The Bacup–Facit section closing in 1947. Although it was now the terminus of one line only, it was quite busy on their visit with ex-LYR 2-4-2Ts Nos. 50647 and 50887, BR 2-6-2T No. 84017, Fowler 0-8-0 No. 49667 and ex-LYR 0-6-0s Nos. 52443 and 52549. The latter was to be the train engine for the return south through Rawtenstall on the 3.54pm to Bury (Bolton Street), the opportunity being taken to photograph the halt at Summerseats then back to Ramsbottom. Here they caught the 4.50pm Salford–Rose Grove handled by Fairburn 2-6-4T No. 42153 as far as Accrington. It was then a train to Blackburn and Hellifield via Daisyfield and Clitheroe to Skipton. This completed the first week of their holiday.

The next morning, they travelled to Keighley, for a trip along the Worth Valley, where Ivatt 2-6-2T No. 41325 was on the 9.15am to Oxenhope. Travelling back as far as

22nd April 1954 • An interesting combination at Skipton on 22nd April 1954 finds LMS 4F 0-6-0 No. 44586 and LNER B16 No. 61413 double-heading a freight.

23rd April 1954 • Ex-Midland Railway Class 2P 4-4-0 No. 40562 stands over a remarkably clean ash-pit at Accrington.

24th April 1954 • After arrival at Clayton West, the loco hauling the 4.10 p.m. from Huddersfield, Black Five No. 44834 can be seen in the distance as it stands at the head of the return train in Clayton West station. Clayton West Station signal box is on the left. Henry Casserley can be seen striding purposefully back towards the goods shed and the station on the right. The residents of Clayton West fought hard to keep their line open, but it eventually succumbed to closure at the beginning of 1983. However, all was not lost as a 15" gauge miniature railway, the Kirklees Light Railway, has been built along the trackbed from Clayton West to Shelley Woodhouse, just short of the former junction with the Huddersfield-Penistone line.

Ingrow, then known as Ingrow West they crossed over to Ingrow East and the ex-GNR line through Cullingworth, Wilsden, Denholme and Thornton to Queensbury and Bradford. I'm not sure about the next stage of their journey but I like to think it was over the GN route from Bradford to Wakefield Westgate via Batley and Dewsbury. A short ride across to Kirkgate and onto Pontefract Monkhill and Castleford Cutsyke. The next journey is on the 2.02pm from Castleford Central to Leeds as far as Stanley via Methley Junction with N1 0-6-2T No. 69440 in charge. From Stanley they travelled back to Leeds behind Class N1 No. 69471 on the 2.26pm from Castleford. Huddersfield was next on the agenda, here they took the 4.10pm to Clayton West, via Berry Brow. Although evening was closing in, they weren't finished for the day and travelled from Huddersfield to Halifax before catching the 6.45pm to Keighley, via Queensbury, with N1 No. 69447 at the head, changing platforms at Keighley they returned to Skipton.

The first photograph taken on the Sunday morning was of the Midland Hotel, presumably the base for the previous three nights, followed by 8F 2-8-0 No. 48105 on a passing freight, 2-6-4T No. 42286 on the 10.00am to Manchester, BR 4-6-0 No. 73045 plus 2P 4-4-0 No. 40414, 4MT 2-6-0 No. 43112, and 4Fs Nos. 43893 and 44197 on the shed. With 2MT 2-6-0 No. 46493 in charge they took the 11.38am to Leeds via Bolton Abbey, Ilkley and Arthington. They arrived in Leeds in time to catch the 1.40 p.m. to Skipton behind Crab 2-6-0 No. 42748. This train was routed via Burley in Wharfedale and Ben Rhydding. Arrival back in Skipton and a change of trains took them over the Settle & Carlisle to the border city and a visit to Kingmoor shed. Caught by the camera on shed that day were 2P No. 40602, Compound 4-4-0 No. 41141, Fowler 2-6-0s Nos. 42802 and 42884, ex-Midland Railway 3F 0-6-0s Nos. 43241, 43301, 43514, 43622 and 43636, 4Fs Nos. 43973 and 44008, Stanier Class 5 4-6-0s Nos. 44726, 45012 and 45281,

25th April 1954 • Ex-Midland Railway Class 3F 0-6-0 No. 43301 basks in the evening sunlight at Carlisle Kingmoor. The loco was one of seven members of the class allocated to Kingmoor at the time. Five were photographed with only Nos. 43351 and 43678 avoiding the camera.

26th April 1954 • By the time Henry caught ex LNWR 'Cauliflower' 0-6-0 No. 58409 on the 26th April 1954 it was one of only a handful of survivors from the 310 strong class built between 1882 and 1902. It is seen at Penrith. Built in 1900 it had a further 20 months of service before withdrawal.

Jubilee No. 45728 *Defiance*, Royal Scot No. 46104 *Scottish Borderer*, Coronation class Pacific No. 46224 *Princess Alexandra*, Fowler 0-8-0 No. 49640, ex-Caledonian Railway 3F 0-6-0Ts Nos. 56317, 56333 and 56355 plus two BR Clan class 4-6-2s Nos. 72006 *Clan Mackenzie* and 72009 *Clan Stewart*.

A photograph of Class J21 No. 65092 on the turntable at Penrith started the next day's adventures, a day which would see them leave the west side of the Pennines. Ivatt 2-6-0 No. 46491 was captured shunting with classmate 46447 in the station on the 10.15 for Workington, which would take the Cockermouth, Keswick and Penrith line on its journey to the coast and ex-LNWR 'Cauliflower' 0-6-0 No. 58409 outside the shed. No. 65092 was to head the 10.30 to Darlington, this was taken as far as Appleby East where a short walk to Appleby West, on the Settle & Carlisle line and a journey south to Garsdale, Hawes Junction as it was once known, where, amongst the subjects caught on camera was the stockaded turntable, this was erected after the wind took control of an engine being turned and spun it round numerous times. Then it was north again to Kirkby Stephen where they again changed stations, this time from West to East, after having captured Royal Scot 4-6-0 No. 46112 *Sherwood Forester* on the up Thames-Clyde Express. Kirkby Stephen East, the western gateway to the line over Stainmore Summit, had once been the territory of ex-North Eastern locos but the only one evident on this day was J25 No. 65655, the other locos recorded were all 2MT 2-6-0s of both Ivatt and Riddles design, namely Nos. 46471, 46474, 46481 and 78017.

It would be No. 46471 that took them forward arriving on the 3.35 p.m. Penrith-Darlington. At Barnard Castle a further two BR 2-6-0s Nos. 78014 and 78015 were photographed during a change of train for a trip along the branch to Middleton in Teesdale behind Class G5 0-4-4T No. 67284. It wasn't just locos they were photographing, station views were taken along the way at Cotherstone, Romaldkirk and Mickleton. The day ended upon arrival at Darlington.

The next day, 27th April, commenced with a visit to Newport shed where A5 No. 69831, B16 No. 61474, J26s Nos. 65744, 65750 and 65766, Q6s Nos. 63341, 63360 and 63446, and T1s Nos. 69910, 69916 and 69917 were photographed. Another of these large 4-8-0Ts, No. 69922, was captured shunting in the yards with two more J26s, Nos. 65733 and 65762 passing on goods trains. Also noted were Class D20 No. 62347 running light engine and L1 2-6-4T No. 67742 on a passenger train. From Newport they made the short journey to Middlesbrough and a train for Whitby, not via the coast as may be expected, but the inland route via a reversal at Battersby. The train engine was Class A8 No. 69861 and on arrival both it, on the 1.22pm to Whitby, and G5 No. 67343 on the 12.00 noon Whitby–Middlesbrough had to manoeuvre around each other to continue their journey. Even today, modern DMU's need to reverse here on their journey between Middlesbrough and Whitby. After capturing photographs of both Kildale and Glaisdale stations they arrived in

27th April 1954 • The roundhouse at Newport evokes memories of the smoky interiors encountered so often on visits to steam sheds. Unfortunately, it isn't smoke that creates this illusion but a grainy negative due to the camera setting required to get this shot in the gloomy interior of Class T1 4-8-0T No. 69910 on 27th April 1954.

Whitby where they photographed A8s Nos. 69858, 69860 and 69880, G5 No. 67302 and J25 No. 65650 and some ex-NER coaches. No. 69861, their train engine from Middlesbrough, was now heading the 4.20pm to Scarborough and took the train up to Whitby West Cliff station where it ran round before heading south. The view from Larpool viaduct as the train left West Cliff was recorded looking both inland and towards Whitby and the sea. They also recorded lineside scenes around Robin Hood's Bay, Ravenscar and Staintondale. At Ravenscar the 4.35pm Scarborough-Whitby was noted with A8 69854 in charge. After arriving through the tunnel into the throat of Scarborough station reversal was again required to access the platforms. One photograph of an ex-GER coach and their loco for the next stage of their journey was all the time they had before Class D49 No. 62766 *The Grafton* set off on the 6.20pm to Hull (Paragon) and an overnight stay.

In 1954 there was still much to photograph in Hull with loco sheds at Alexandra Dock, Botanic Gardens, Dairycoates and Springhead plus passenger workings out of Paragon station and the goods lines circumnavigating the city centre. However, they headed for the railway owned station that never saw a passenger train, Hull Victoria Pier, and the ferry across to New Holland on the south bank of the Humber. Knowing the area reasonably well during the 1950s and 1960s I like to think they walked from their hotel through the city centre to Queen Victoria Square and along the dockside, Princes Dock and Humber Dock, to the pier. The alternative was to catch the Pier bus, service 50, from the bus station.

The Man from the Pru will conclude in Issue 3 with the final two days of the journey on 28th and 29th April 1954, accompanied by tables showing the stations, halts and loco sheds photographed during the journey.

27th April 1954 • With the famous Whitby Abbey dominating the skyline there can never be any doubt to the location of this photograph. Class A8 No. 69880 is seen adjacent to the engine shed. The River Esk, which splits Whitby in half is in the background, between the railway and the abbey.

These two pages contain bonus images from the collection purchased by Paul King, who we thank for allowing the reproduction of them in the pages of Eastern Times.

25th August 1950 • Henry Casserley in the cab of ex-GER Tram Loco, Class J70 0-6-0T No. 68217 at Upwell on the Wisbech and Upwell Tramway. The vans would be carrying fruit to the various markets around the country.

13th May 1936 • The negatives I wanted were often included in larger batches which brought bonuses such as this. A little under eight months after entering service, the second built Class A4, No. 2510 *Quicksilver*, is seen at Newcastle at the head of the Silver Jubilee, for which service it was built. Note that is still carries the recessed coupling hook which was removed after a tragic accident at Kings Cross when a person was trapped and seriously injured whilst coupling up a member of the class.

4th September 1954 • Class C14 4-4-2T No. 67440 is seen at Macclesfield Central between journeys from/to Manchester London Road. Not all negatives are perfect and this had a blemish in the top left corner where I have done some repair work.

WEST HARTLEPOOL (51C)
IN PICTURES

The Stockton and Hartlepool Railway, which connected the town of West Hartlepool with the Clarence Railway near Billingham, was opened for goods on 12th November 1839 and to passengers later the same year on 1st December. A station named Hartlepool West was opened on 9th February 1841 which was then renamed West Hartlepool in February 1848, closing on 3rd May 1880 when it was replaced by a new West Hartlepool station. This in turn was renamed Hartlepool on 26th April 1967 when West Hartlepool was merged with Hartlepool following the complete closure in 1964 of the former Hartlepool Dock and railway station in the Headland, this station being previously known as Hartlepool. The shed was built and opened by the North Eastern Railway in 1867 closing 100 years later in September 1967. Peppercorn Class K1 2-6-0 No. 62045 is seen at her home shed in 1966.
Photo: Wally Cooper (WC1136) © The Transport Treasury

Gresley Class A3 4-6-2 No. 60073 *St. Gatien* makes a spirited departure from West Hartlepool with a northbound express service.
Photo: Neville Stead Collection (NS204820) © The Transport Treasury

Gresley Class V3 2-6-2T No. 67690 at West Hartlepool on 10th September 1953. *Photo: Neville Stead Collection (NS204816) © The Transport Treasury*

Thompson Class B1 4-6-0 No. 61013 *Topi* arrives at West Hartlepool
with an express working having passed the elevated Church Street Signal Box.
Photo: Neville Stead Collection (NS204809) © *The Transport Treasury*

York (North) allocated Peppercorn Class A1 4-6-2
No. 60141 *Abbotsford* at West Hartlepool in 1963/64.
Carrying the headcode for a Fish/Perishables XP working it looks
as though there's a spot of shunting in progress before heading south.
Photo: Neville Stead Collection (NS206637) © *The Transport Treasury*

Raven Class Q6 0-8-0 No. 63435 pictured inside West Hartlepool shed in the early 1960s. These engines hauled heavy mineral trains on the North Eastern Railway. Built by Armstrong Whitworth the loco pictured was in service from May 1920 until withdrawn from 51C in June 1966. © *The Transport Treasury*

W. Wordsell designed Class J73 0-6-0Ts Nos. 68358 and 68364 stand outside West Hartlepool shed in this undated image. *Photo: Neville Stead Collection (NS202199)* © *The Transport Treasury*

THE EDINBURGH FAST

BY IAN LAMB

Just like the real thing, some of the Class 47 Co-Co diesel electric locomotives are still running today after almost sixty years of service, like my Hornby model [R060], renumbered and painted to suit my preference.

The class entered traffic in 1962, and by 1968, 512 examples had been built. At one time or other they could be seen virtually everywhere from the Scottish Highlands to Penzance, and from Holyhead to Great Yarmouth, hauling all manner of trains ranging from express passenger to a humble ballast working.

Most of my rail travelling days were behind these locomotives all over main line Scotland – indeed I once had the privilege of being on 'the footplate' of the 'push-pull' unit on a journey from Aberdeen to Glasgow appropriately named *The Duke of Edinburgh's Award*. This locomotive was modelled by Hornby [R887] in the late eighties and can still be obtained on the 'swap-meet'/second-hand stands. Whilst I really enjoyed viewing the line ahead rather than from a side window, undoubtedly my favourite journey behind a Class 47 was over fifty years ago.

Whether one thinks of the Forth and Tay Bridges, or the twisting sinew of a line across the 'Howe of Fife', this county has always presented a challenge for railway operation. Nothing has ever been quick or straightforward in 'The Kingdom', or has it? Certainly it is not the ideal terrain over which to catch up on lost time, but one can have a damn good try. That is exactly what a determined railwayman tried to do on a very cold winter's day on 15th February 1973.

"Dunday, Dunday. This is Dunday. British Rail wish to apologise for the late running of the 09.05 train to Edinburgh." This was too good an opportunity to miss whilst I stood on Dundee Tay Bridge station platform so I feverishly fumbled through my brief case for a note pad and pen, and with watch at the ready prepared to enjoy what the local newspaper – 'Courier and Advertiser' – called *The Anatomy of a Remarkable Rail Run* when they kindly featured my article a short while afterwards.

Not a Class 47 but two BRCW Type 2 locos, with D5385 leading, double-heading a train past Wormit signal box after departure from Dundee across the Tay Bridge on an unrecorded date. Photo: Sandy Murdoch (SM015-12) © The Transport Treasury

30th June 1973 • Class 47 diesel-electric locomotive No. D1975 passes through the former Leuchars junction with what is believed to be an Aberdeen to Edinburgh express. The bay platform once housed the local service branch to St. Andrews. *Photo: W.A.C. Smith (WS10019) © The Transport Treasury*

The 7.45am 'up' Aberdeen to Edinburgh express (09.05 ex-Dundee) had unfortunately been delayed owing to locomotive failure prior to departure from the 'Granite City'. It was almost 45 minutes behind schedule when it emerged from Dock Street tunnel at the eastern end of Dundee Tay Bridge station. Headed by Class 47, 2580hp Sulzer locomotive, BR number D1515, it was controlled by driver D. Clark and secondman A. Rose, of Ferryhill, Aberdeen.

Considering the various connections this train had with others south of Edinburgh, it seemed likely that great efforts would be made to catch up on lost time. And what a run it turned out to be.

It should be remembered that this route is one of the most difficult stretches of main line in the country because of numerous speed restrictions and the fact that the locomotives have to start and continue their trains on steep gradients, and in some cases very lengthy ones.

With barely time to pack the many passengers into the eight coach train, D1515 departed at 9.48. It roared on to the 1 in 66 incline at the station throat accepting the challenge from this relatively short, but gruelling climb up the curving viaduct leading to the Tay Bridge.

The load of 270 tons emerged from the station yards into a bright warm sun reflecting silverly on the cold, grey Tay estuary with distant Fife shimmering in a haze. Once on the bridge proper the 'Edinburgh Fast' appeared to coast through the arches with the brickwork of the old bridge piers standing out in sharp relief against the dull and hazy background.

Turning sharp right at the old Wormit station in good time, it accelerated immediately and steadily on the 2¼ miles climb to disused St. Fort station, then burrowed under the recently constructed Tay Road Bridge approaches with the track of the former Tayport branch veering sharply to the right.

Slowing down for the Leuchars curve was inevitable. Then, with "whistle wide", we rushed across Leuchars station at precisely ten o'clock. Fast now through the relatively level Eden Valley and frosted Fife, like a knife cutting cleanly through an iced cake, until compulsory slowing at the former British Sugar Corporation sidings on approach to the Cupar curve and station at 10.06am.

Along the almost straight and level track through Stratheden speed increased steadily. With horn blaring, and braking, D1515 took the Ladybank curve and junction in its stride at 10.12am beneath the shroud of the snow-capped Lomond Hills. At maximum effort, it thundered over the Falkland Road Bank and through the Freuchie Gorge, passing Markinch, Haig's Whisky and all at 10.17am.

Braking was delayed until absolutely necessary on approach to Thornton Junction over the speed restriction due to subsidence risk from the old coalfields. However, though straining like an animal on a lead, speed had

25th August 1972 • A Class 47 diesel-electric locomotive is carefully controlled round the fierce curve beyond the station at Burntisland with an Edinburgh bound express. Photo: W.A.C. Smith (WS9794) © The Transport Treasury

dropped drastically in slowing down at Thornton. Once on the Dysart descent, we ran freely past Frances Colliery sidings, and swayed on the sharp curve leading into Sinclairtown.

Roaring through Kirkcaldy station at precisely 10.25am, D1515 produced increasing speed on the level past Stark's Park and Seafield Colliery. On to Kinghorn, braking hard on approach to the station (10.28); observing the maximum 30mph speed allowed at the tunnel, wheels screeching on the curve.

Accelerating hard to the allowable 50mph along the 1 in 305 embankment above Burntisland sand flats and sheltered beach, braking hard again for the notorious and fierce Burntisland 'S' curve above the town thoroughfare at 10.32am with the wheels taking it rough.

From a mere crawl past the British Aluminium complex, its red oxide dramatically off-setting the glittering frost, we accelerated hard. A 50mph limit was like chains to this effervescent locomotive, straining at the leash to be up and away. Aberdour at 10.36am.

Now clear of basic restrictions, we headed fast towards the inevitable severe curve at Inverkeithing station (45mph limit) at 10.40am. Speed gradually increased on the 1 in 70 incline to the Forth Bridge approach, throbbing and vibrating vigorously through the tunnels.

We emerged to the vista of the magnificent bridges at North Queensferry at exactly 10.44am. Then we faced 40mph speed restrictions across the cantilevers before building up for the fast run necessary through the Edinburgh suburbs. Silvers and greys were the prominent colours in the mist and haze over the Forth embankments.

Speed increased from Dalmeny (10.46) then, with a challenging roar and impertinent whistle D1515 was away, leaving the right-hand branch to meander its way to the west. We ploughed through the mist engulfing Turnhouse Airport and clattered past Saughton Junction (10.51) before receiving the first noticed caution on the journey whilst approaching Haymarket sheds for the run in to Edinburgh Waverley.

A slow crawl under The Mound at 15mph, then we were gliding into Platform 11 and stopping abruptly at 10.57am. It had taken 69 minutes for the 59 miles from Dundee, seven minutes less than scheduled timing.

It's a fact, although it seems barely credible, that in August 1895 – at the time of The Great Railway Race – the East Coast train was worked over the 59.2 miles from Edinburgh

13th April 1963 • The Forth Bridge and ferries viewed from South Queensferry. *Photo: W.A.C. Smith (WS6626) © The Transport Treasury*

For the moment the Forth Bridge has disappeared in mist as an eventual Class 47, currently No. 1560 accelerates through Dalmeny station with an Edinburgh bound train. *Photo: W.A.C. Smith (WS9885) © The Transport Treasury*

No. 47053 on a Dundee-Edinburgh service is seen at the one-time Saughton Junction on the western outskirts of Edinburgh shortly before the signal box was demolished, and the track realigned, to give no crossing facilities until Haymarket West. *Photo: Ian Lamb*

to Dundee in the record time of 59 minutes. There's little doubt serious risks must have been taken in excessive speed over many curves to achieve it.

In the days of my youth spent happily on the walls of Saughton Junction, when A4s and V2s on the Dundee run used to race the Glasgow expresses, I doubt if any of these locomotives in their heyday could have matched the performance of D1515 and the skill of her crew.

Yet, perhaps it was nostalgia. For I could have sworn there was the ghost of Gresley in that engine, such was its achievement on this restrictive route – and on reputedly the coldest day of the year. I may have been slightly late for a business appointment, but boy, what a run!

Even now, when I think of that journey, my blood still tingles at the memory of urging the engine on; trying hard to keep my cup of coffee from falling over. Nevertheless I was so proud of that locomotive that I sought to have its efforts truly recognised so managed to establish that it was based at Gateshead.

Therefore, being an English allocated machine (I desperately hoped it would be a Haymarket one) I wrote to the railway authorities in York suggesting that in honour of its achievement it should be named, ideally *Bonnie Dundee* – not least of all after my favourite A1 class steam locomotive of that name (No. 60159) which regularly ran over Fife metals.

I got a really snooty letter back informing me that it was not the custom or the intention of that region to name diesel locomotives (at a time when the Western Region were doing just that?) so I left the matter there. Apart from renumbering my Hornby model at the time as D1515, as far as I am aware there is no official acknowledgement of that locomotive's great achievement.

On reflection I am sorry that I didn't follow up the named suggestion. Perhaps had it been a Scottish based machine then I might have been more successful because around a decade later Haymarket allocated 47/7s – especially modified for 'push-pull' operation – were appropriately named.

2nd August 1958 • 'A1' Pacific No.60159 Bonnie Dundee coasts down from the Forth Bridge with the 6.45pm Edinburgh to Dundee Tay Bridge through Inverkeithing South junction. Photo: W.A.C. Smith (WS253) © The Transport Treasury

THE HEADSHUNT

In future issues our aim is to bring you many differing articles about the LNER, its constituent companies and the Eastern and North Eastern regions of British Railways. We hope to have gone some way to achieving this in Issues 1 and 2.

Eastern Times welcomes constructive comment from readers either by way of additional information on subjects already published or suggestions for new topics that you would like to see addressed. The size and diversity of the LNER, due to it being comprised of many different companies each with their differing ways of operating, shows the complexity of the subject and we will endeavour to be as accurate as possible but would appreciate any comments to the contrary.

We want to use this final page – The Headshunt – as your platform for comment and discussion so please feel free to send your comments to: tteasterntimes@gmail.com or write to Eastern Times, Transport Treasury Publishing Ltd., 16 Highworth Close, High Wycombe HP13 7PJ.

Sir

I recently purchased a copy of Eastern Times Issue No. 1, Winter 2023 and am finding it most enjoyable. The photograph on page 37 (left) in the article on Concrete Signals is Luton Hoo station. The old station has been retained as a private dwelling.

In Jauary 2011 whilst walking on the Lea Valley Way I found the concrete signal post lying in the undergrowth alongside the remains of the station platform. I doubt whether it is still there as Sustrans have done work to improve the old trackbed since then.

If you are interested I have a digital photograph of this LNER concrete post taken at the time (see below).

Regards,
Michael Harding